501 Ways to Crimeproof Your Home

Protection from crime and natural disasters

by Karl Kratz

PREFACE

Can you visualize someone right now breaking into your home? Are you prepared? Do you know what to do? This informative book can instruct you how to protect you, your family and your possessions from falling victim to a criminal act or natural disaster. You must become responsible for making your home secure.

Myths & Falsehoods

You may think it can't happen to you or you live in a safe neighborhood but it happens all the time. You may think you are protected by your dog; however, you are still vulnerable. You may think you don't have anything worth stealing, but everyone has some valuables. You might have insurance but it wont cover your life. You may think if it's going to happen, its going to happen and there isn't much you can do, however, there is plenty you can do, and you might prevent it. If you do not prevent it, at least your actions can make recovery from crime more manageable. You may think the police can handle it; they are the experts but there are usually only two police officers on duty for every 1,000 citizens including 5 ex-convicts. Police unfortunately stress apprehension after the fact and not prevention or victim's rights.

Why Read This Book?

Police officers cannot be everywhere. Over 40 million crimes are committed each year. There are 6 million burglaries yearly with more than 1 billion dollars worth of goods reported stolen. Everyday there is a burglary every 10 seconds; nearly 50% are committed without force. No window is unbreakable. A person who makes a living getting into people's homes can always find a way. You cannot completely eliminate crime or the professional burglar; however, by physical and psychological deterrents listed in this book, you can substantially reduce your chances of becoming a victim. The objective of the burglar is to get in, to get possessions and to get out without being seen. By removing the objective, you can prevent crime. Criminals are lazy and avoid tough jobs when they can find an easier target: you will learn how not to be an easy target.

This book helps reduce worry as well as help protect you, your family and your possessions. You can minimize the risk of becoming a victim of a murder, a burglary or an assault. The best protection against crime is to be prepared for it and to anticipate it. Putting preparation off could cost you. Take action by reading this book.. A 1972 study by the Department of Justice stated, "The most important recommendation that we can make is that the ordinary citizen realize that, by a series of simple straightforward acts, he can affect the likelihood of being burglarized."

ACKNOWLEDGEMENT

I am deeply indebted to Cal and Penney Lum and Sherry Bussey, they are close friends who supplied needed editorial talent. I also want to thank my brother Ken Kratz for his help in the formation of this book.

I gratefully acknowledge the assistance provided by the following during the research stage of this project: Crime Prevention Center, California Office of the Attorney General, John K. Van de Kamp, 1515 K street, Suite 383, Sacramento Calif. 95814; U.S Department of Agriculture; Consumer Information Center of the General Services Administration; U.S. Department of Health and Human Services; National Center for the Prevention and Control of Rape; U.S. Department of Labor Employment and Training Administration; U.S. Department of Justice; National Institute of Justice; National Criminal Justice Reference Service and Technology Assessment Program; U.S. Consumer Product Safety Commission; California Seismic Safety Commission; Federal Emergency Management Agency; Federal Bureau of Investigation; National Commission on Fire Prevention; National Fire Protection Association; American Red Cross; Middletown Ohio Division of Police; Pima County Attorney Victim Witness Program, Tucson, Arizona; Seattle Washington Police Department; San Jose Police Department, California; Fund for Modern Courts, Inc., New York; Santa Clara Valley Water District, California; Pacific Gas and Electric Company and Pacific Bell. The information and public domain pamphlets that they provided as a public service were very helpful in the formulation of this book.

Statistics were taken from the Federal Bureau of Investigation and the United States Department of Justice reports and other national agencies who have the never ending task of making sure we live in a safe environment while insuring our right to all of our liberties.

NOTICE: The author of this workbook attempted to produce reliable and practical advice. This book intends to relate the experiences of crime victims and outline some of the actions they took to deter crime. We cannot, however, guarantee that its application will safeguard people or property. No claim is made by the author that the steps outlined will work with criminal encounters in different situations. Each individual should seek professional services as required to cover his specific requirements. While the author made every effort to insure this book's accuracy, the burden is on the reader to check and evaluate all statements. Liability for any losses that may occur in preparing for, during or following any emergencies as a result of using the guidance in this publication is specifically disclaimed.

You can order more copies of 501 Ways to Crimeproof Your Home by mailing $14.95 in a check or money order to INNPRO.

INNPRO

P.O. Box 2079

Los Gatos, Ca 95031-2079

501 Ways to Crimeproof Your Home

Published by INNPRO

ISBN 0-941201-02-3

Library of Congress Catalog Card Number
88-083285

Table of Contents

Chapter 1

INTRODUCTION TO SECURITY

What this Book is About

This book presents many straight forward, easy to follow steps to reduce the likelihood of you, your family or your possessions falling victim to criminal acts and natural disasters. Many ideas are tips you will want to know and others will reaffirm your common sense. The book has few hypothetical stories or pages of statistics, but instead, mostly good advice written for the average person. Read the whole book and not just sections. If you skip a section, you might miss valuable recommendations that pertains to more than one type of crime. If you don't have the time to read the whole book, read the security checklist and look over the rest. This book is for the entire family; people from high school age to grandparents should read this book.

Steps to Protection

1) Recognize crime for it does happen. Be aware of crime; know where, how and why it happens.

2) Deter crime both physically and mentally. Make it difficult for criminals. Make it hard for criminals to get in. Use locks, alarms, stickers and common sense to stop accessability. Make it less profitable for criminals; make it hard for them to find items and remove valuables. Don't carry cash or keep expensive valuables in the home. Mark all valuables and use a safety deposit box. Make it life threatening for crime; make it likely he will get caught.

3) Know the proper response when encountering crime. Know what to do during a natural disaster.

4) Limit losses; buy insurance and use the included inventory sheets.

5) Recover and make corrections; make sure the crime doesn't repeat itself. Take self defense classes. Buy a better alarm. Report and prosecute criminals.

Who Commits Crime

Most likely a burglar is the amateur, age 15-22, who uses drugs. The second most likely burglar is the semi-pro, age 23-30, who comes day or night and will dress as a delivery man. He usually carries weapons and will fight if provoked. The least likely criminal you will encounter is the professional who does not carry weapons and has a high degree of skill in avoiding alarms and police. Unless you have a great deal of very expensive valuables (e.g., furs, art work or diamonds), the professional will probably avoid your house. A professional burglar is the only one you stand little chance of warding off, but don't worry because he is a very small minority.

About Crime

Crime is not only an separate incident. A burglary can escalate to far more serious crimes such as mugging, rape, assault and even murder. You should not fear crime and violence; nevertheless, you should be concerned about it. With the rising number of women in the work force, there has been a increase of break-ins in the daylight hours. Prime time for criminal activities are the months of July and August. These months have higher incidents of crime due to the fact that more people are on vacation. Other likely times for crime to occur are Saturdays, the Christmas season, New Year's Day and during a fire or natural disaster requiring the attention of the authorities.

Disclaimer

Each criminal encounter is unique. Each situation is unique. Some things might work and others may not for your situation. This is a book of options that have worked in the past but might not be the best in situations that you come across. Only you can be the judge of the situation.

Chapter 2

PROTECTION AROUND THE OUTSIDE OF THE HOME

General Area Outside

A medium level of security is recommended for residential security. The various levels are: maximum with a sophisticated alarms, cameras and armed guards; medium with an alarm, a dog and high security locks; low with good locks and physical barriers; and minimum with simple locks and simple barriers.

Burglars cruise neighborhoods and judge each house. They start with the outside appearance of your home. The burglar's primary considerations are: places to hide while gaining entry, the proximity of other people, quick escape routes, types of doors and windows on the premise and the visibility of his actions. He will search for the weakest link. You can discourage a burglar by not providing the "perfect environment."

Doors

Good doors are solid core which means the door is of solid wood at least 1 3/4 inches thick. A steel door with inside hinges is better than a weak veneer door with a hollow core. A good locksmith will charge an average of $350 - $600 to install a dead bolt and a solid core door. Don't purchase the locks that spin free, latch type locks, or locks with easy to pry off knobs. Fix any gaps over 1/8 inch wide in a door hinge. See illustrations in the back of this book. You can boost protection and save energy by placing a heavy metal guard strip/astagal to the door edge if there is a gap between the door and frame or between the doors in the case of paired doors.

Install a peep hole in your main door so you can view whoever is there before you open the door. A one-way viewer that lets you see out without outsiders seeing in is available. A convex mirror opposite a peep hole is a good idea for blind spots.

Tamper-resistant hinge screws should be used for doors with outside hinges. Also, for outside door hinges, you should install a security pin. Sliding doors and windows can have added protection with nails and screws. Sliding glass doors are especially vulnerable; always use more than one locking device to secure them. Sliding doors should have vertical dead bolts and broom handles to go in the bottom of the door's frame.

Use double cylinder locks instead of single cylinder locks on most outside doors. Doors with glass that is within 40" of the lock should have double cylinder locks. Double cylinder locks can only be opened with keys; for this reason, keep the keys handy if you use these types of locks. Caution: double cylinder locks are restricted in some communities because this creates a exit/entrance barrier during a fire. Check with the local fire department before installing them.

Garage doors should also be solid core. It is a good idea to keep garage doors locked. If a burglar can get into your attached garage, he can probably work there unseen until he breaks into your house and he probably will use your tools to do this. Use a 5 pin tumbler padlock with a 3/8" shackle and along with a strong hasp.

Exterior Lighting

Leave at least one light on all night long. Don't let outside lights burn all day long because it is a tip-off that your house is unoccupied. Never leave just the outside lights on because this also makes your house look vacant. If you use regular incandescent bulbs for outdoor light bulbs, they should be only 15-25 watts to prevent them from breaking when they get wet. It is best to have at least a 40 watt bulb in front and a 60 watt bulb in the backyard and shield them or buy special outside bulbs.

It is best to have the lighting wired to a timer, photo cell or any other type of device that comes on automatically. You should light all entry points and report broken street lights since criminals sometimes disable lights so they can do their work unseen.

Landscaping

You should plant hedges; even short ones act as a barrier that delineates what you consider as yours. Keep your hedges well trimmed and low enough so passers-by will be able to see an intruder trying to break into your home. Shrubbery and bushes that can hide a person near an entry or window should be trimmed or removed. Planting cactus, pyracantha or any thorny plant near openings to buildings will help prevent culprits from lying in wait. Branches that make it easy for a thief to climb into a second or third story window should be cut off. Fencing is good for delineating property. It should be in good repair without openings in or under it and should not offer criminals a place to work unobserved.

Locks

Remember, valuables left unlocked have no protection. If your locks are over five years old, think about updating them. If your locks were installed with economy and not security in mind, have them replaced also. Have your locks replaced if they have any of the following symptoms: the lock is loose on the door; you have to wiggle the outside knob in order to turn or depress locking buttons on the inside; you have to slam lift or push down on the door in order for it to stay shut and locked; you have trouble removing your key from the lock; you have to lift, push down or wiggle the key around before it will turn to lock or unlock the door.

There are new rotating pin locks that offer much greater protection. They have far more keying combinations. The new pin tumbler locks are more precise than disc or wafer types. Most lock set replacements are surprisingly easy to install.

For doors, minimum security is a dead bolt with at least 1" throw (bolt extending at least 1" from the door) and a matching strike plate with 3" case hardened screws. Rim mounted locks with dead bolts also provide good security and are easy to install. Use shim resistant locks. I do not think pick proof locks are worth their high cost. Avoid spring latch locks for they are easy to shim open in under ten seconds with just an ordinary credit card. Heavy chain lock all entrances but do not count on these locks as your only defense for they can be easily broken by even an average size man.

Select locks that are best suited for the particular security problem. There are special locks for every type of door, windows, garage cellars, patios, bikes and recreational vehicles. There are special locks for window ventilation that offer some protection and also allow you to get some fresh air.

Have your building owner put up a grill at the foot of the back of stairs if there isn't one already there. If your apartment building has a master key, change the lock or add an extra lock (see if it is okay with the building owner first). When you move into a new home, replace the locks for you never know how many or what type of people have keys from the previous occupants. Insurance is not a replacement for good locks; nevertheless, acquire both. See lock illustrations in the back of this book.

Windows

No household window is unbreakable, but all your perimeter entry ways should be protected at least minimally. All windows should have secondary locks added to them if they only have the simple catches provided by the builder or manufacture. If your windows just have a small thumb turn lock, realize they can be pried open or easily opened through a broken pane. Add a key lock and make sure everyone knows where the key is in case it is needed for an emergency exit.

Shutters are nice for all windows. They offer protection and add value to your home. Sash windows should be protected with pins. If you use iron window grates or grills, make sure they can open from the inside because these can trap you in a fire.

If you must open your windows frequently for ventilation, make sure they can lock when in the ventilation position. If you have a choice, open windows from the top not the bottom. Opening a window from the top makes more noise and is more difficult to open when burglars try to gain entry. Nailing the top of a window down on your air conditioner helps prevent people from stealing your air conditioner or removing it to gain entry. During a party, remember to place guest purses and coats in a room where the windows are securely fastened.

Put grates over basement windows. Basement windows and garage windows should have curtains. Basement windows that are used for ventilation should have ventilation locks. For glass windows that are in isolated areas, think about replacing them with hard to break plastic type windows. Plastic windows may be worth the cost. See illustrations of windows in the back of this book.

Dogs and Pets

Dogs are excellent for increasing security. Often they are even better then an alarm if you have the space and don't mind the inconvenience. Barking dogs warn and draw attention but they are not a replacement for locks for animals are less reliable. I suggest you choose a puppy and raise it to keep as a friend and a burglar deterrent. All breeds of dogs are good for security. Even small dogs are a warning device. Junk yard dogs and attack trained dogs are sometimes dangerous to have as household pets for they can attack a neighbor or your child and could put you, the owner, in serious trouble with the law. In addition you may also be sued since you may be responsible for your dogs actions. Most cities have leash laws; don't let your dog loose on your property unless you have a fence or your dog is leashed.

Professional dog training has advantages and disadvantages. It is expensive and dogs must be retrained periodically. Most dogs do not need to be trained to increase your security. It is an instinct for dogs to bark at strangers, but you should train them not to take food from strangers since this stops their barking. Teach your dog to bark and not growl. Give your dog the run of the house. Don't punish dogs for barking at strangers as someday you might need the noise. Feed your dog in the morning and don't let strangers feed your pet. It is best to have just one or two members of the family feed the dog so that the dog won't get accustomed to taking food from many people. If you install a doggy door, make sure it is not a way for a burglar to get in.

Attack dog signs can help even if you don't have a dog; however, most burglars know about this trick. A large dog dish or dog chain placed in the back yard can have the same effect as owning a dog. If you live in the country, you might consider geese instead of a dog for security. They are territorial minded and make good guards. Ancient Romans even used geese because of their keen sight and hearing as well as their loud warning honk.

Good Habits

Remove temptation, items that are left out are a target for petty theft. Keep valuables out of sight and out of reach. Don't keep house keys, car keys and your address together. If you lose them it will be a lot harder to sleep at night. Don't leave messages on your doors, like "I'll be back in two hours." It is like leaving a notice to a burglar saying, "Please rob me." Don't leave a ladder around and lock your gates. It is not wise to leave your trash can outside all day after a pickup, newspapers on the lawn, or mail in the mailbox, because this action gives criminals a clue that no one is home. It is best to park cars in the garage where they are safe. A lot of burglars are not deterred if cars are parked out front. Never leave a bicycle outside or at school overnight. There are thieves who specialize in stealing bicycles. Watch out for quick working thieves while you are in the back yard doing chores; this is a prime time for burglars to enter your house from the front and quickly carry off your valuables. Be cautious of building storage areas. They are not very secure and have a high amount of thefts. Be conscious that people could be following your routine. Instead of always leaving the living room light on when you go out, try to be unpredictable, (e.g., leave the radio on sometimes). Amateur burglars many times live in the same neighborhood and learn your habits so vary your routine. For instance, walk your dog at different times and at different places every day and start varying other daily responsibilities as well. If there are strangers in your neighborhood, ask them politely what they are doing. This also goes for strangers in your building lobby. If a car repeatedly goes around the block, get its license number. If these strangers do not have bad intentions, there is no harm done. I would not depend on people employed by private security companies for they are usually under-paid, under-trained and not well regulated.

One Time Precautions

Know your neighbors on all sides and tell them to call police and then call you if they think there is trouble. Tell them that you will do the same for them. Neighbors should know each others family's living habits - who comes and goes. This will make strangers easier to spot. If you see a stranger who appears to be doing something out of the ordinary, call the police immediately. Do not keep a spare key outside. Leave a key with a trusted neighbor instead. All professional criminals, including inexperienced ones know where to look for hidden keys. Decals on your door should not show the make of alarm if you use one. Lock all skylights or have them alarmed. Many burglars use open skylights as easily as you use a door. Never display guns where they can be seen through an outside window. Use a mail slot instead of mail box so you won't build up uncollected mail and samples left by salesmen. Place Operation I.D. warning stickers on doors and windows. Add extra initials to your mailbox if you are living alone; you do not have to let strangers know you are alone.

List only your street address outside. Do not list your name. Fancy signs that say "This is the Jones' residence" are a great help to burglars. Have only your address clearly marked and lit for emergency vehicles especially when there are older people living at the residence. Address signs are best if numbers are 6" high and are a contrasting color or are made of reflective material. If your house is on a corner, post numbers on the street you use as the address. If your house is some distance from the road, post your address near the road. If your local police uses a helicopter patrol, painting your number on your back roof with characters 24" high in a contrasting color will enable them to locate you quickly. Many police departments have seminars on home security. Phone and ask if you may attend one or put one on yourself. The Citizen Watch Program and others like it offered by police are well worth organizing. The police say it reduces crime 45% in some instances. Watch out for developing a false sense of security with the program. You can make it more of a block awareness committee to avoid citizen watch burnout. You are not good enough to do it alone, so get friends and neighbors involved. I am sure your apartment will be more secure if all the people in your building look out for one another by writing down license plates and shouting or blowing whistles when they sense trouble. Burglars will realize they are dealing with an entire building and will leave it alone. The program is easy, just use your eyes and ears and then phone police. Before police departments were formed in the large eastern seaboard communities, citizens conducted street patrols themselves. You should always call the police when you sense trouble. Don't try to stop a criminal yourself.

Chapter 3

PROTECTION INSIDE THE HOUSE

Do You Need an Alarm and Where Should You Buy It?

Why buy an alarm? Alarms help to discourage crime and will give you peace of mind. Burglar alarms have proven to be an effective tool in the detection of crime. You should decide if you need an alarm by looking at the type of neighborhood you live in and your predilections. If you feel vulnerable, you probably are. You might have to buy an alarm in order to know that you are doing all that is humanly possible to protect yourself and family. Consider both the cost and the nuisance of alarms.

I recommend an alarm if you keep extremely valuable property, live in a isolated area or simply want more protection. Alarms work 24 hours a day wether you are at home or away. Alarms can both decrease or increase damage. If it enrages a burglar on drugs, it could make things worse. If you live alone, I recommend, at the very least, you buy an alarm for your bedroom window.

Cost and Benefits of an Alarm

A Portland, Oregon study on burglar alarms showed that people without alarms had 6 times as many burglaries. However, be aware that high priced alarms don't always equate to more protection. The extra gimmicks could mean more complexity, more failures, and less convenience. Remember, over protection does little good and is money wasted. Also, if you buy an alarm, I suggest you don't skimp on price either. If you skimp, you will probably get many false alarms and this extra nuisance is not worth the added savings. The average price of an alarm is between $300 and $1,000. The average burglary nets about $600; therefore, this small risk doesn't justify thousands of dollars worth of alarm equipment for most people.

Alarms may decrease the cost of your insurance substantially and may be tax deductible. Check with your insurance company and accountant. Alarms can also add to the resale value of your home. The amount you spend depends on the dollar value of the possessions you want to protect. Don't let alarm companies play on your fears after a burglary. There are many unscrupulous salesmen that take advantage of human weaknesses and fears at this time.

General Information About Alarms

When you get an alarm, also get instructions on how it works and how to maintain it. Tell neighboring adults what to do if they hear the alarm, who they should call and how to shut it off. Alarms are not toys so never abuse them. Remember the story of the boy who cried wolf.

Types and Selection of Alarms

There are two basic types of alarms. The first is an area alarm and its name implies just what it is used for; it protects spacious areas. Area alarms work on various principals; for instance, infrared light, ultrasonic waves, heat, light, microwave, motion and sound activation are available. These are activated when someone enters a protected space. Perimeter alarms are the second type and, like the name implies, they protect perimeters. Perimeter alarms can be hard wired or wireless; they activate when someone tries to get through the perimeter. There are also combination alarms available that use both types of detection. There are even closed circuit TVs used for crime prevention. Spot alarms are another type that cover just a particular spot like an alarm located under a carpet or attached to a painting.

Determining the best alarm for your residence is resolved by analyzing your needs. Both type of alarm systems have their advantages and disadvantages. Perimeter alarms are the most prevalent type in the home for it sounds before burglars get in and they are less likely to have false alarms. You, your pets and your children can run freely inside your residence with your alarm armed and ready to detect intrusion. I think this is a great convenience. Area alarms are used especially when it is hard to place sensors around the perimeter. For example, if you have an unusually large number of windows, they are worth considering. When choosing an alarm, also consider ease of installation, placement and the appearance of the alarm.

The alarm system you choose should have a backup power supply, installation and maintenance procedures, a panic button near the door or by your bed as well as a panic button near the control box. The alarm sound should be loud enough to be heard inside and outside of the house. Some alarms can be connected to an autodialer that dials a friend, alarm company or the police. When an alarm dials to an alarm company, it is called a central station hookup. Contact police to see if they offer this service and ask for their opinion. Most police departments are against auto-dialers because it ties up the phone lines during a major disaster. Remember, alarms are only warning devices. Alarms do not initiate action. Someone must be able to respond to the alarm for it to be effective. You should not make your home a fortress. The goal to keep in mind is to delay burglars and make it worth their while to go to other places; the goal is not to downgrade your quality of life.

Hiding Places

There are many places to hide valuables in your home. Included are: refrigerators, washers, dryers, other major appliances, under baseboards, inside stairs, in clothing, under chairs, in ceilings, in hollowed out books or fake wall sockets. The places you should never place valuables are under the bed, inside a medicine cabinet or in night stands and dressers.

At the Door

Use a peep hole first, then a heavy chain when identifying strangers, but do not rely on your chain exclusively. Don't open the door to people stating emergencies. Instead, make the calls for them. They may say, "May I use your phone? My car broke down." or "I am lost. Could you give directions?" But instead they might be wanting to get inside to rob you. If their claims are legitimate, they won't mind you making the call for them.

Do not assume a uniformed person and people making door to door appointments are legitimate. It is best to talk to their employer. Verify delivery men, building and fire inspectors by calling a number from the phone book and not by the number the visitor provides. If they run when you place the call, you should call the police at once. Ask to see a permit or valid identification or credentials from all door to door sales people. If a door to door salesman does not have credentials, call the police.

Never let a stranger know you are alone in the house. Don't let anyone in that you don't know or haven't called. When alone, and before answering the door at night, you may first want to call out something like "I'll get the door, Brian." Many rapes and robberies occur in the victim's home. In addition, many burglars assess your valuables and security measures contemplating a future break-in while at your door .

On the Phone

If you receive a phone call that turns out to be a wrong number, ask what number they were dialing and just say they dialed wrong. Never state when you will be home to strangers and don't reveal personal information to strangers over the phone. Give this advice to your children also. Don't tell your phone number to people you don't know well. Be aware of people who seek information about others and verify their reasons for asking. Callers who hang up without talking might be burglars checking to see if anyone is home. Use this situation as a warning signal to check your security.

Report obscene phone callers, and I recommend you blow a whistle in their ear, but some authorities say they phone back and blow a whistle back at you. If an obscene caller keeps harassing you, write down the date and times of the calls and use it as proof in order to convince the phone company to set up a trap. This action costs the phone company money so they are reluctant to do this, but talk to the supervisor and their criminal division and be insistent. I think this is better than you having to change your number. It is not submitting to a sick person.

I recommend you get a phone answering machine in order to screen callers. Use your first name only on a phone answering machine's recorded message tape and give the impression you are at home but cannot come to the phone.

Inside Lights

Leave some lights on all night long and use lights when you move from room to room. Use timers on indoor lights or a radio when you leave the house. Also, leaving a bedroom light on is more believable then a kitchen light.

Safes

If you buy the right type of safe it will be good protection against unskilled burglars, fire, rushed thieves, and kleptomaniacs, but no safe is safe from a professional criminal. Watch out when buying small safes, for they can be carried out by burglars and opened at their leisure. Only some safes offer fire protection and some are better than others. When considering the purchase of a safe, it is best to shop around and read their ratings to find one that meets your needs. Avoid placing a floor safe or container in the bedroom for it is the first place checked by burglars. Use a safe deposit box at a bank. The cost may be tax deductible; check with your accountant. Store expensive small items in a safe deposit box even if you have a safe at home.

Safe deposit boxes are a good place for a marriage certificate, divorce and adoption papers, citizenship records, service papers, birth certificates, deeds, auto titles, house hold inventory sheet, stocks and bonds and contracts. Leave wills with a lawyer and not in a safe deposit box because courts seal them and it might take months to get them open.

Security Closets

I recommend you make a secondary barrier also called a security closet. It is used as a retreat or safe haven in case of an attack. The closet should be the master bedroom or its walk-in closet. It should be secured with a key lock with a solid core or steel door and contain a phone and a weapon. You can also use the security room to store valuables like furs, guns, silverware and cameras. You can also make a security room in your basement or garage. Make a good sized room out of studs. Cover the wall studs with 1/2 inch plywood nailed every 5 inches. Use 1/2 inch type X wallboard for fireproofing and mount a small louvered vent high on the wall. The door should be at least 1 inch thick and be solid core. Inside the closet mount a light fixture, an alarm and store your safety supplies.

Confronting a Burglar

First of all, try not to encounter a burglar. If possible, get out of the house as quickly as you can. If you confront a burglar, don't panic, but instead cooperate because the money is not as important as your life. Say "I'll give you what you want if you will leave." Chances are they will drop what they are doing and just run away at this point. The saying is not threatening and it puts "leave" in their minds. Criminals live with fear and have little or no sympathy for their victims. Never struggle with a burglar unless it is your last resort and you are clearly in danger of physical harm.

When at home and a burglar enters, never try to stop him yourself; every burglar has the potential to kill. If you hear a burglar, get to a phone and quietly call the police if you can. Try to stay on the phone until police arrive. If you cannot get to a phone, make a loud noise and chances are you will scare them off. If you walk in on a burglar in your house, pretend you think he is a repairman until he leaves. If you hear him inside your bedroom at night when you are in bed, you should fake sleeping until it is safe. Remember to say "I'LL GIVE YOU WHAT YOU WANT IF YOU LEAVE."

Set-Ups

Set-ups are people who finger you as a target for a criminal. These can be many people including but not limited to: the gardener, garbage collector, mailman, paper boy, newspaper or magazine subscriptions solicitor, appraiser, your insurance man, Avon lady or any door to door salesman, etc.. Be aware of anything unusual about their services and don't volunteer information about vacation plans or living habits to them. You can even set yourself up if you advertise a lost dog or disclose future vacation plans with a listing of your address.

Good Habits

Deposit your checks soon after receiving them. Do not get on an elevator with a suspicious stranger; wait for the next car. Don't ride the elevator going up when you want down, but instead wait for it to return down. Do not go to the laundry room alone; criminals know it is frequented by women, and it is isolated with many noises to cover up screams. Close your blinds and drapes at night. If you are leaving, remember to always use your locks even when you are just leaving for a minute. Leave cheap decoy jewels for criminals to take. Leave your radio or T.V. on low volume tuned to a talk show. Leave the phone off the hook sometimes. Don't leave large amounts of cash in the house, and lock your check book up. When you are coming home, approach the door with your keys in hand.

One Time Precautions

Check to make sure that valuables are not visible from the street since they are tempting to intruders. Keep in mind that most burglars are not professionals, but instead are probably kids from the neighborhood. Keep a table near your door entrance to set packages on when returning from shopping which enables you to shut the door right away. Know where the nearest public phone is located. Get an unlisted phone number if you are single or especially worried about crime.

General Do's & Don'ts in the House

In an emergency, give your address first to the ambulance people over the phone. When you place an ad in a newspaper use your work phone. When you advertise a lost pet, a car for sale, a garage sale, etc., make sure to lock up and check your security. Watch workmen when they come into the house. If two workmen come in to fix something, tell them they have to stay together and tell you where they are at all times. If you are expecting a delivery, notify your building attendant. Report all crimes to police for it is always a help. Criminals develop favorite working areas and ways of doing things so all the information is of great help to police officers in apprehending criminals. Check the complete list in the back of this book for items to store in the house in case of an emergency.

Chapter 4

PROTECTION ON THE STREET

Guns and Lethal Weapons

Should you carry a lethal weapon? It is not a good idea because it is probably illegal; laws differ from state to state. It can be unnecessarily dangerous and it forces violence into situations that a lot of times can be avoided. Should you own a gun? Anti-gun factions claim guns are six times more likely to accidentally kill its owner, member of family or friend than it is to repel a burglar. Don't believe that a gun in your apartment is the answer to all your security needs. If you own or plan on buying a gun for your home, take a gun safety class. They are offered at junior colleges and sporting good stores. Know the gun laws and don't shoot unless you are in real danger of being harmed. Remember guns can be taken from you and used against you, so don't even pull it out unless you are capable and willing to use it. Always keep guns in good repair and stored unloaded with a trigger lock. When a gun is needed, it must be loaded and accessible; an unloaded and locked gun is neither. Unfortunately more guns are stolen each year from homes than burglars are captured by home owners at gun point.

I have experienced both the positive and negative sides of gun ownership. I have sold guns that have stopped robberies which, under the circumstances, could not have been prevented any other way. I have also had a very close friend fatally shoot himself when going through a depression. I only recommend gun ownership if you take the time to learn how to use it properly and are a responsible person who won't reach for the gun to settle an argument. The best gun for home protection is a pump shotgun. It makes a distinct sound when loading a shell that all burglars easily recognize. Take classes in order to learn how and when to shoot. Bullets and pellets can easily pass through a wall and hit other people.

Tear Gas

Tear gas sometimes just makes an attacker mad, but some police departments recommend it for most people as the best thing to stop a street crime. There are different types; Cn takes 2-3 seconds and Cs is stronger but must hit the attacker's skin to be effective. They are illegal in some states and in other places you have to take a class and become registered to buy or carry it. Talk to your local police for more information if you are considering purchasing it. I live in California where a person must receive training to become certified to carry tear gas. I have taken the training and I believe tear gas is a effective deterrent to attacks if you are aware of its limitations.

Nonlethal and Everyday Weapons

Every day weapons include an air horn or a sound device, rat tail brush, hat pin, nail file, ballpoint pen, keys (key to the groin), cork screw, umbrella, water pistol with ammonia, hair spray, lit cigarette or lighter to the eye, salt and pepper. All of these have been used to ward off an attack, but some have made things worse by infuriating the attacker. There is no way of knowing if they will be effective in any situation you come across.

For protection, I highly recommend the air horn or sound device for the sound attracts the attention of others and the device cannot harm you if taken from you. Try it out when you buy one so you won't be scared of the sound it makes. I do not recommend the broad casting of the spice pepper for I think it is not effective. If you are going to carry any of these weapons when entering a dangerous area, walk with them in your hand ready to be used. Learn how to use weapons on attackers weak areas and use them with gusto. If it makes you feel more secure, sleep with some weapon near your bed on the nightstand (i.e., scissors or at least a paper weight to throw to break a glass window).

Self Defense Fighting

I recommend all types of self defense training. Self defense classes help you display self confidence in street and are a lot better than books for building self confidence. You are the only judge in a situation as to whether or not to fight. It depends on you, the attacker and the whole situation. Some people recommend fighting if the attacker doesn't have a weapon or if he wants to take you somewhere. Without knowing the situation, no one can tell you what is best.

If you decide to fight, try to surprise the criminal. Fight with commitment. Scream and go for assailant's weakest spots. Don't be squeamish. Poke your assailant in the eyes with your thumbs. Knee him in the groin. If you are grabbed from behind, step your heel into his instep and give him your back elbow to his stomach. If you are outnumbered, submit by dropping to the ground and draw knees up while placing hands behind your head. This action will probably reduce your injuries. Also, talking to the attacker might give you the chance to get away. Fake fainting has proven to be useful also. Remember, yelling "fire" can sometimes get a bigger response than yelling "help." Be cautious, yelling fire inside a building may cause panic. Many street criminals view victims as an enemy and derive pleasure by inflicting injury.

Confronting a Street Criminal

Criminals look for easy targets to go after. Like lions stalking herds of animals, criminals detect fear and look for weaknesses. If you are approached by a criminal, he has already sized you up as a weaker adversary. As discussed earlier, the decision to fight or not is yours alone. There are many confrontation options (i.e., fight, run, negotiate, response or be unresponsive). If they have a gun, look them in the eye not at the weapon; it will help you maintain your composure. If possible, avoid being taken into an alley or driven to a remote area. If the attacker is armed and insistent, it is probably best not to fight. Avoid making any quick moves. An attacker is usually as nervous as you are and probably more jumpy. Try to cooperate with a mugger unless physically attacked. If physically attacked, try to get away and, if you cannot get away then fight.

If you choose to fight, fight dirty. Muster strength. Take a deep breath, think logically, get composure then go for it. If attacked in an elevator, press as many buttons as possible. Try to avoid all criminal confrontations; if you return and there is someone inside your home or you see a sign of a break-in, go to your neighbor and call police and wait outside for them to arrive. If you see your belongings at a flea market, get the police. Don't try to get them back yourself.

If you see a mugging, start shouting and urge others to join in. We must get mad and join others if we are to rid the streets of crime. There are many street punks who get their reward from the knowledge that others fear them. If we ban together, we can stop them.

Be leery of any individual who is acting irrational (i.e., someone who gets angry over the right to a parking place). People have been murdered just for standing in a line. Don't be a one man army. Avoid dealing with irrational people for they cannot reason. Try to shrug them off.

When confronted by street hoods, be natural and walk right past them. Don't stare, and ignore their remarks. If they bump you trying to incite trouble, keep walking. They know exactly what they are doing and have probably done it many times before. Remember they are in a superior position if you are by yourself.

If Followed by a Stranger

If followed by anyone, speed up and switch directions; this lets them know they are spotted. Never go directly home when followed. Instead, walk until you see help. Walk to any occupied house or lock yourself in any open car you can find. To escape, run into the middle of the street if it is safe from traffic and yell. If approached, look for other people or a lighted window. You can make a phone booth an emergency sanctuary just by sitting down with your feet against the closed booth door. If possible drop your billfold in a mailbox. Waiting around the next day for the mail pickup is better then losing your wallet to a mugger.

Safety Precautions: Do's and Don'ts on the Street

Things to avoid and things that warrant special precautions include: people in parked cars, doorways, shrubbery, public transportation, beaches, parks, school campuses, playgrounds at night, alleys, bushes, deserted areas, trash and garbage bins, public rest rooms, laundromats, 24 hr. teller machines and being alone on the street anytime especially at night (night crimes are 3 times more frequent). Staying in lit areas is not as important as some people say for criminals are human and need some light also.

If watched when coming out of your house, pretend to have a friend back inside the house. Yell back at the house, "Brian, I'll be back in five minutes." This will make the people watching you think you have a friend that can hear you.

Carrying a whistle is a good idea, but don't wear it around your neck where it could be used to strangle you. Look around first before using an automatic teller machine. Stand at the teller machine sideways and put the money you get out of it in your front pocket. If you are at an automatic teller and see a suspicious person, pretend you forgot your number and leave. It is best to go with a friend to a teller machine if you have to go during off hours. Don't short cut through isolated areas.

If a car pulls up to you or goes by several times and a stranger asks directions, stand away from the vehicle when you help them. Avoid groups of teenagers on the street. It is wise to cross the street if a group of males approaches and you are by yourself. When in parks, don't go off to explore unknown areas by yourself. Traveling in groups on the street is safest and more fun. Always stay in well populated areas. Jog with a partner or in groups and always stay together. At crowded gatherings such as outdoor concerts, stay with friends. When sitting on a lawn picnicking or at a concert, keep your possessions in front of you where you can see them.

There is a lot of crime at bars and night clubs because people's senses are down and criminals know this. Keep a small amount of cash in a separate pocket to pay your bar tab. Don't show a bankroll. Watch where you walk and never walk alone when under the influence of drugs or alcohol.

Carrying Valuables

It is best to carry as little cash as possible. I recommend you carry only $40 in small denominations and divide your valuables in several pockets. Place cash amounts, keys, and your ID in separate pockets. Keep your keys in your pockets and not in your purse. This prevents the thief from getting into your house after he has stolen your purse. If you must carry large amounts of cash, divide it and carry it in separate pockets. You should carry spare money in pockets to make a phone call in the event that you lose your wallet. Only carry one credit card. When you carry multiple credit cards, carry them separately and only carry them if you are planing to use them. Put a quarter in your wallet's photo pocket to make a call in an emergency. Sew hidden pockets in clothes for cash- especially for elderly. In public don't show gold chains, cash and credit cards, and turn rings around so stones cannot be seen. If you make large bank deposits for a business, don't take your own valuables along with you, and consider getting an armored car service. Make several deposits a day to limit losses and vary your route and schedule so a robber will never know when or where to expect you.

Public Streets

If you sense something is wrong, then something probably is wrong. If you feel vulnerable, then you probably are vulnerable. Watch out! Be cautious of strangers who seem inappropriately interested in you. If your purse becomes lost or stolen with your house keys, replace or rekey your locks. Change your locks whenever you lose your keys and ID. Don't talk about valuables in public places this will just invite trouble. If you see a crime or a non-injury accident, phone to get qualified help. Only give personal attention in an accident if there is no danger to yourself and there is no one already helping. I highly recommend a CPR class. Cardiopulmonary Resuscitation Classes are offered at hospitals around the country they will train you to save lives in an emergency. Watch your luggage in the no-man's land located between the taxi and the hotel lobby. Arrange instant deposit of pay checks, social security, and all other incoming checks. If you take away the criminals opportunity, you will reduce your chance of becoming a victim.

Public Transit

When riding busses, don't sit near the exit door or in the back of the bus. People can snatch your valuables when the bus pulls up to a stop; rowdies congregate in the back of buses away from the driver. Avoid isolated bus stops. During off hours, ride near the driver and, if you have trouble with people, tell the driver. When waiting for a bus, stand back from the curb so you cannot be grabbed from a passing vehicle. If you must wait for a bus at night, back off from the street so you cannot be seen until the bus comes or wait in a well lit area like a restaurant or a gas station. At a bus stop be alert; being stationary makes you a target. When in a bus terminal, have your back to a wall. When stepping off a bus, look around to spot trouble before it happens. If there is someone who looks like trouble, stay in the open. After getting off a bus, don't walk down isolated streets. Instead look at your watch and act like someone is going to pick you up. Don't open a purse or wallet while boarding transit. It is best to have your money ready in hand to pay the fare.

Never use general public transportation when intoxicated. Try to avoid sleeping on public transit. Stay out of general public bathrooms at all terminals for they are very dangerous. When standing, place your briefcase between your legs touching you.

When taking a subway, sit in a populated car and avoid the last car. Don't stand near the edge of the subway platform. Watch the people when there are only a few people around. During off hours, stand near the token booth and outside the turnstile if possible. Don't walk down dangerous corridors alone; wait for other people. If there are trouble makers aboard, don't stare back at them; move to another car. Stand near the exit if you cannot get away from them. If they assault you, pull the emergency break cord. This will stop the car to summon help and make it easier to catch the criminals.

Use a taxi whenever possible and don't get into an unregistered taxi. Don't use a taxi if there is someone with the driver. In most places it is against city regulations since the additional passenger may be there to rob you.

Other Good Habits

When you are given a lift home, have the driver wait until you are inside before having him drive off. When coming home, take your keys out in the lobby of your building to have them ready when you reach your front door. Have your keys in your hand ready for the door so you are not stalled at the door fumbling for them.

Use the elevators instead of the stairs in buildings and ride an elevator only in the direction you intend to travel. When waiting for an elevator to pick you up, stand back to make it harder for someone to grab you. While in an elevator, stand near the control box. Beware; pickpockets use crowded elevators to practice their trade. If a stranger gets off the elevator with you, make sure he doesn't follow you to your apartment. While you are in phone booths, face the street and don't even use the booth unless it is well lit.

When walking on the street, make eye contact but don't stare at people or walk close to the curb. Walk assertively and face traffic. Keep your handbags in front and close to your body. What was true for you as a child is still true for you as an adult; don't talk to strangers.

Pick Pockets

If you have ever misplaced or lost your wallet or purse, you know the awful feeling. Much of this horror can be eliminated by planning ahead. If you live in a very high crime area, carry a fake wallet with $10 in it to use as a decoy. Carry only a few business cards in your wallet because you don't want someone to masquerade using your identification. Don't carry your wallet in your back pocket; instead, carry your wallet inside your coat. By placing a comb horizontally inside or wrapping rubber bands around your wallet you make it much harder to steal without you knowing. Be especially cautious in airline and hotel lobbies, subways, ball games, concerts, flea markets and all crowded areas.

If someone bumps into you, check for your wallet. Your drivers license, credit cards, bank cash cards, business cards, etc. are worth a lot of money to criminals; therefore, contact police immediately if they are stolen. If you want to make a claim against your insurance you will probably need to file a police report anyway. When you lose your wallet, notify all your credit card companies.

Keep your own master list of credit card companies up to date in the inventory section of this book. When you need a new license, go into the Department of Motor Vehicles with proof of your identity, (i.e., social security card) and get a new license for it is illegal to drive without one. If you carry a blank check in your wallet, you might have to close the account since many banks won't let you stop payment on a check when the amount is unknown.

Purse Snatchers

A recent Seattle study showed that 80% of purse snatchings were reported from women walking alone; for this reason, shop with a friend. Staying alert is about the best thing you can do. Don't carry a purse if you can avoid it; use your pockets instead. Don't fight purse snatchers for they can be violent. Your purse can be replaced, but your life cannot. Contrary to popular belief, there are professional purse snatchers; in fact, criminals have even coined a term for them: "rippers." Carry your hand bag like a football (snug to your body under one arm) and don't let the purse strap go across your body. If you are right handed, carry it under your left arm. Keep your best arm free for emergencies. Do not wrap purse straps around your wrist; people have been injured during purse snatchings because their purse became entangled. While you are walking on the street, carry your purse or brief case on the side farthest from curb. Some people like to carry their purses upside down with their hand on the zipper so when a thief pulls it the contents are spilled all over the ground. If you live in a high crime area, think about buying a purse alarm that sounds when snatchers pull your purse out of your hands. Most thieves will drop your purse rather than run away with it if it is giving off a siren alarm. Keep your purse off bathroom floors and don't hang it on the top hook on the stall door. Hang it on the second hook instead. When you are in theaters, keep your purse on your lap. If your purse is snatched, look around for it might be dumped in the trash after the thief removes your cash.

Shopping

Dress for shopping; don't wear high heels, fancy clothes or expensive jewelry. Make an itinerary before you leave. Shop with a friend and buy expensive items last or arrange to have them delivered if possible. When making a purchase, don't show large amounts of cash or many credit cards. If you write out a check and they ask for a phone number, use your work phone and not your home phone especially if your checks have your address imprinted on them. Don't talk about personal matters to shop clerks (i.e., telling your butcher your vacation plans or the dress shop about your new fur or TV). Keep an eye on your purse when it is in a shopping cart. You should always destroy all your carbons from your credit card transactions. Watch out for set-ups of people who take your packages just after you paid for them. While shopping, don't overload yourself with bundles. Try to keep at least one hand free. Store packages in your car trunk as you collect them instead of carrying them around while shopping. Look around before you place them inside your trunk and place them in quickly. Try not to let people see you load your trunk. If you don't have a trunk, place packages on the floor and not on the highly visible seat. Be especially cautious in your entryway when you return from shopping.

Security at the Office

Try to avoid working alone. If you work after hours, lock your office doors and turn on a radio or TV. When finished, walk to your car with security personnel or a friend. Vary your regular routes. Unpredictability will make you a harder target. Never leave your purse or billfold in plain view or in a coat hanging on the door. Use operation ID on personal office items (instructions are in the back of this book). It is best not to leave valuables at the office. Always report suspicious persons to the proper authorities (i.e., management, security or law enforcement). Keep emergency numbers by the phone and know the best emergency exits.

Chapter 5

PREVENTING VEHICLE THEFT

Driving Precautions

Approach your car with your keys in hand. Before you get in, walk around the car and check the back seat for criminals. Routinely check the back seat after parking on the street or in a garage. Always lock all your car doors after entering and before leaving. It is a good idea to drive with all your doors locked. Always lock your car even if you are getting out for just a quick stop. Windows should be rolled up and locked. Close all windows including the little vents when parked. Don't roll down your windows until you leave a parking garage. If you own a convertible, keep the top up if you are driving alone. At a traffic light keep your car in gear. Try to keep your gas tank at least half full. While traveling, keep your purse on the floor of the car, not on the seat next to you. Never leave your vehicle with the motor running. Many people who go to quick stop markets and leave their engine running while they go in to buy cigarettes or milk find out car thieves are faster than cashiers.

Car Alarms and Locks

I recommend you protect your second largest investment with an alarm. It may also reduce your auto insurance rates. The auto alarm dealers I spoke to said most alarm buyers are people who have just suffered a major loss. Don't wait; it might be your turn next. A million cars are stolen each year and less than half are returned.

There is no way to stop a professional car thief. Some crooks have even used cranes to pick up cars and put them on trucks. However, you can protect your car by discouraging car thieves. Four out of five cars stolen are left unlocked and one out of five have keys in the ignition. Most cars are still taken by amateurs.

Car alarm systems come in many different types. There are locking, disabling and alarm security devices commercially available. They work off motion, gravity, sound, voltage changes or any combination of these types. Think about getting an alarm with a panic button if you feel insecure in your car. Alarms should have a siren or horn and use lights to get attention. Sirens on alarms are illegal in most jurisdictions if they sound too much like police sirens. Some alarms can also offer protection against accessory thefts. Silent alarms for average vehicles are too expensive and, therefore, not cost effective.

Locks for the whole car include ignition replacement locks that cost about $150 which are better than the manufacture's original equipment. There are also cuff locks that go over the ignition and steering column, and there are cane bar type locks that connect the gas pedal with the steering wheel. There are many criminals who take only the expensive items out of cars. Some people call them "car boosters" or "car clouts", in any case, protect your valuable items with special locks. There are many other types of locks that frustrate thieves; for instance, there are interior hood lock releases to protect your engine, locking gas caps, locks for batteries, boat engines, wheels, spare tires and tape decks just to mention a few.

Methods for disabling vehicles include elaborate devices such as installing fuel cut offs and ignition interrupt systems (i.e., a fuse mounted inside the coil which stops the electrical spark) as well as simple schemes such as removing the rotor from the distributor. All of the above are called kill switches and are effective.

It is a good idea to install headless door lock buttons in your vehicle and to install an automatic garage door opener for added security. Buy a multi-frequency garage door opener so that it won't be accidentally opened when a plane flies overhead. Periodically check your garage door, if it has an opener to make sure it is properly adjusted so that it cannot be lifted up enough to let a criminal crawl under it.

Buying and Selling a Used Car

Before buying a used car, always check with the Department of Motor Vehicles in your state for the actual owner of the vehicle. Be suspicious of a low asking price, fresh paint, replaced locks, no address of seller, out of state stickers, older plates on a newer car and replaced glass. Inspect the vehicle identification number for alteration. Any of the above mentioned conditions could mean the car is stolen. If you buy a stolen car, you will probably end up losing your money. When you sell your car to a private party, always ask to see a valid driver's license. Ride along on test drives and be cautious. If you are a woman, bring a friend. Don't take a personal check; only a certified cashier's check won't bounce.

Accident Procedures & Disabled Vehicles

The following steps listed should be followed in the event of an accident (listed in order of priority): call the police, call the insurance company, call your auto club and then call a relative or the person who is expecting you or who can help you. If you see a disabled motorist, go to a phone and make the call for them after helping unaided injured.

Many times it is best not to get out of your car on the road at night even after you are involved in a accident. If no one is hurt in a nighttime accident, agree with the other driver to drive to a lit area and then report the accident. Don't get out in the dark (caution: in some states it is illegal to leave the scene of an accident; check with your local police and your insurance agent). Don't be flagged down by anyone except police and beware of people who will cause a minor accident in order to rob you.

If your car breaks down, open your car's hood and attach a white cloth to your radio antenna. If you don't have an antenna, attach a white cloth to the drivers side door handle or attach it to a window or your windshield. There are windshield sun screens that have "Get Help" printed on them that are good if you do not put the wrong side out by mistake. Afterwards, get back in your car and roll up the windows and lock the doors. When someone stops, crack open your window and tell them to call the police or a garage. If the car becomes disabled while traveling in a remote area, everyone should go together and leave a note saying who you are, where you are going and when you left. If you are a single woman, you may not want to show that you have become disabled; for this reason just wait for a police officer and flag him down. If you are disabled at night, you may want to stay in your car until sunrise. When you get a flat tire on a busy road, keep driving slowly till you find a safe place to pull over to change your tire.

Special Vehicles

When you park campers, trailers and large recreational vehicles, you should remove portable valuables. Park them in a clear open space, remove the distributor rotor and keep an inventory list of what is inside. Have your drivers license number painted on your campers roof in large letters so police can spot it from the air if it is stolen. Always chain and padlock all trailers, motorcycles and snowmobiles even when they are stored in a garage. It is best to register bikes and boats and always pad lock them with a case hardened plastic covered chain of at least 3/8 inch diameter. Lock bikes so that both the back wheel and the frame are locked to a stationary object. Remember to always lock bikes even at home.

Precautions: Do's and Don'ts for Vehicle Owners

Keep your car in good running order. Read your car owner's manual and follow its service recommendations. Have duplicate keys made. Using duplicates reduces the chance of having them copied from the numbers on the keys. Keep your vehicle's spare key in your house not under the vehicle's hood or wheel well. Use Operation I.D. markings on cars, car parts and car items such as tape decks. Record your vehicle identification number which is on a plate on the dashboard near the windshield. Don't even drive through the seedy parts of town.

Crime and Personal Vehicles

If your car won't start it may have been intentionally disabled by a criminal so don't take unsolicited offers of help. Any time a stranger looks like trouble and is hanging around, sound your horn continuously until he leaves. If you are followed while in your car, drive in the center lane to a police station rather than head for home. If your car appears broken into, get assistance right away because the criminal might still be in the area or watching you. If you see anyone beneath or near your car, go back the way you came and get help. If taunted by rowdy gangs, don't think of your isolated car as a safe haven; instead, seek populated areas. I know it is hard to do but don't let bad drivers upset you, for it can easily lead to a fight. When in danger, keep all the car doors locked and honk your car's horn on and off to get attention.

General Motorist Tips

A little planning and prevention techniques can go a long way in deterring crime. Some safety tips include: not traveling at night, using well traveled roads even when there is a short cut another way and not picking up hitchhikers. Watch your gas station attendant when he checks the oil, water and tires. Join an auto club. Remember, docks are high crime areas and should be avoided. If you live in a unpopulated area, buy a citizen band radio (CB) or an emergency radio that you can use to call for help. If your car is stolen, time is of the essence. Notify the police at once. After the police come, call the insurance company, you can panic later.

Parking

Parking lots are smorgasbords for criminals. There are cars, accessories, packages and people for the taking. Look around when getting out of your car to spot trouble before it happens. When parking, don't get out of your vehicle if you see any suspicious people. Make sure you lock up everything before you leave the vehicle.

Turning your wheels sharply inside to the curb will make your car much harder to tow away. Be cautious in underground parking. If you are returning to your car by yourself, wait for an escort or talk loudly as if going to the car with a friend. It is usually better to park on the street at night rather than enter an unattended underground apartment garage. When opening your garage door at night, keep your car's headlights on and pointed inside your garage to make sure no one is hiding there. It is best to park your car facing the street if you are worried about people tampering with your engine since the vandals will be easier to see.

Your car is 5 times more likely to be stolen at night if left in an unattended lot. Make sure you park in an attended lot when going to a concert or other publicized event because criminals do a lot a cruising during these time periods. Knowing how long you will be away makes things easier for criminals. Give a parking attendant only your ignition key. If he gets your house key, he can make a copy of it, get your home address from your registration and rob you later.

If you are storing your car for a long time, remove your registration and any other papers with your identification from the glove box, take your parking voucher with you, write down the odometer milage to prevent parking attendants from taking joy rides, remove the distributor rotor, remove all easy to carry items (water skis, camping gear) and conceal accessories like car stereos. Do not tell the parking lot attendant how long you'll be gone.

Chapter 6

SEXUAL ASSAULT

General Rape Prevention

Federal reports vary from one in every three women to one in ten women will be raped or assaulted in their lifetime. You probably will not be raped, but you will be safer if you read this section. Rape is a crime of violence in which sex is a weapon. It affects all age groups and nationalities. It can happen in seemingly safe places. Approximately a third of all rapes happen in or near the victims home. Rapists are not always strangers. In over one half of reported rapes the rapist was an acquaintance, neighbor, friend or relative of the victim. The intensified public awareness has increased the reporting of rapes committed by dates and family, especially by Hispanic and Oriental women. Oriental women have been reluctant to report rapes due to their cultural background. Most rapes occur outside; so read the "On The Streets Chapter" as well as this section. Be aware that a date could turn into to rape. As a precaution have disguised telephone code words planned with roommates and friends meaning, "I am in trouble; come quickly" (make it sound like you don't want to be interrupted). Supervise school age dates and enforce a reasonable curfew.

It is a myth that women are raped because of the way they dress; however, scale down on looking too sexy in public. For instance don't wear a see-through blouse without a bra. You don't want to stand out. Wear shoes and clothing that give you freedom of movement since it is very hard to run away in dress shoes and tight skirts. Don't place your bed near a window if possible. If you have printed stationery and are single, don't put your name on the outside envelope.

About 12% of all women who reported being raped were hitchhiking at the time of the assault. Rapes occur more often in the rapist's car than in the victim's. Accept rides only from women or older couples. Ask for the driver's destination before you accept a ride. A lit cigarette is a good deterrent to attack. If you are being driven home and the driver deviates from your destination, stay calm and try to get him to stop the car in order to get something to eat or use a rest room. If he stops, ask others for help. Escape any way you can. In nearly half of the reported rapes only the attacker was under the influence of drugs or alcohol and in 20% of reported rapes both the attacker and victim were under the influence.

It is important to never believe you are always safe; on the contrary, you should always be wary. There are many women's groups like the YWCA, rape crisis clinics and junior colleges that offer educational programs on rape defense and protection. I recommend taking classes to help build your self confidence and awareness.

Confronting a Rapist

Each criminal encounter is unique. In any situation something might work once but not another time. This chapter looks at options that have worked in the past, but some of these options may not be the best choice in the situation that you become involved in.

To fight or not is your decision alone. There are several tactics which might help when confronted by a rapist. Don't be afraid to leave your own home if you are in danger. Building up a rapist's ego might give him enough satisfaction so that he won't get violently physical. Another tactic is to take away the rapist's incentive by saying convincingly you have a venereal disease like AIDs, you are pregnant, going through your menstrual period or are mentally retarded. Vomiting, faking a heart attack and relieving yourself are other ploys.

If the rapist drops his guard, run. If you cannot get away, it is better to be raped than violently harmed. Remember to use self defense, to run and to scream if it is possible, so that someone will hear you. If the attacker is armed, it may be best not to resist. If he is unarmed, maybe fighting will give you enough time to get away.

When attacked, fighting is probably best. While fighting, get evidence; gouge with your nails, cut him, pull his hair and use surprise if possible. If a single attacker asks you to put his penis in your mouth, ask if you can massage his testicles then bite as hard as you can and rip his testicles off. Don't be squeamish and have no pity for him; he has none for you. This will give him enough pain and make it easy for you to get away. Remember, escape is the objective. If you decide to submit, close your eyes and tell the attacker you won't be able to identify him later but definitely report it. If you believe you might get hurt defending yourself or if you are afraid to fight back, don't fight. Submitting to a rape out of fear for your safety does not mean you consented; it is still a rape and should be reported. Do not feel guilty; the rapist is the criminal and not you. Due to fear and shame, many rapes go unreported.

After the Assault

The steps to take after being assaulted (in order) are: get away from the attacker, report the crime to the police, learn corrective measures and receive counseling.

Call police immediately and call a rape crisis hot line. Ask the operator for the number if you do not see one listed; look under "Rape", "Women", "Sexual Assault", "Crisis", "Victim", etc.). A rape crisis hotline can be called day or night. They can explain your options, offer counseling and contact the police if you haven't already done so. After being raped don't change clothes, rinse mouth, bathe or douche until you report to the police because you may be destroying evidence that will help the police apprehend and convict the rapist. Take along a change of clothing when you go to the police station or the hospital so you can give the police the clothes you were attacked in. Your clothes might be needed as additional evidence. Write down the entire incident (see after crime chapter). Be examined by a physician right away and show the doctor all your injuries; make sure your injuries are documented. Be tested for pregnancy and venereal disease get another examination two weeks later.

If you have been psychologically affected, don't be ashamed to seek counseling. If you know or suspect you know of a victim, stay close to them and offer moral support. It is important to report and prosecute these criminals. Getting them off the street will make you feel better and probably save many women and possibly yourself again from going through this violent criminal act. Remember the law is on your side; they are not judging you. Do not feel guilty or ashamed. It is normal to feel fear, anger, loneliness and helplessness after being raped. It is normal to cry. You have the right to be loved; you have done nothing wrong.

Chapter 7

FIRE AND NATURAL DISASTER

Fire Risk

According to the National Fire Protection Association, fire presently kills approximately 5,500 people yearly and injures over 20,000, many of which are children and the elderly. The death rate for the elderly (over 65) and children (under 5) is three times that of the general population. Eighty percent of fatal fires occur between 8:OOpm and 8:OOam because, when you sleep, your senses are less effective. According to the Federal Management Agency, careless smokers account for the number one cause of fire; over half the fires are attributed to them. Next highest are faulty or improper heating systems, which cause 15% of all fires. Electrical equipment and cooking stoves account for just under 15% of all fires but are more frequent than arson. Most deaths are caused by the fumes of carbon monoxide gas from smoke which displaces the oxygen. It is the fumes and not the flames that are the most dangerous.

Smoke Detectors

Your fire warning system must warn you early enough for escape, be dependable and be always alert. Place smoke detectors between potential fire sources and the room where you sleep; this is usually somewhere near the bedroom door. Do not place smoke detectors near vents or heaters. Every home should have at least two and there should be one on every level of your home. The warning alarm for most smoke detectors is audible for 60 feet square. Heat detectors are good for furnaces and fire places, but are not good enough by themselves for personal fire protection. Slow burning fires can cause enough gases to bring about unconsciousness and death before flames trigger a heat detector.

There are basically two types of smoke alarms. One type is the ionization type which has a minute amount of radio-active material (smaller then 1/2% of what you receive from the normal environment). This material changes the air inside the detector into a conductor of electrical current. When smoke enters the chamber, the particles mix with ionized air (combustion gives off electrical charges to make ionized air) which in turn reduces the current flow and sets off the alarm. The ionization type is the simplest in design, has low false alarms, works faster for fumes, is usually battery powered and needs no visible smoke to set it off. They are easily serviced by periodic battery replacement.

The last type of smoke detector is the photoelectric which has a light sensitive cell (the sensor) and a built in light source that directs a beam into a chamber. The alarm is set off when smoke particles entering the chamber reflect the beam onto the cell. To service, test and replace the battery and light source every 3-5 years. The best type of detector operates by reflecting the beam whenever smoke reaches a density of only 2-4%. The photoelectric type can give off false alarms, but it works faster on smoldering fires.

Fire Extinguishers

Extinguishers usually operate by first removing the pin, pointing the nozzle at the fire and squeezing the handle. Read the instructions when you buy one before you have to use it. Use an extinguisher only on small fires and, if you cannot put out the fire right away, get out of the building. There are basically three fire extinguisher ratings for use on the various types of fires. The ratings are: A (for combustibles), B (for grease and paint thinners), and C (for live electrical which requires dry chemical to put out). There are special types of extinguishers like ammonium phosphate that can put out all three types of fires.

Fire Plan

Do your planning now, so when a fire happens, you will know exactly what to do. Your fire plan should include two unobstructed exits from every room. Rearrange your bedrooms so the children have the easiest exits. Install a hall door if it will help. A hall door can close off a living room fire and will allow people to share the remaining emergency exit. If you have keyed locks, have keys handy. Use safety ladders for every second story window especially if it is over nine feet to the ground (store it under the bed). Have an outside meeting place picked out in advance; such as a tree. Do not use a fire plug. Choose someone to report the fire and someone to get each young child out (e.g. an older child can be in charge of a younger one). Many child casualties are found under beds and in closets trying to hide from a fire. Have a surprise practice drill to test your escape plan at least once a year and preferably twice a year. Set off the smoke detector so the children will recognize it's sound. Fire drills are as important for your family at home as for children at school.

Train the whole family to stay near the floor and crouch low below the smoke. Learn to test the door by feeling it with your hand before opening it. If the door feels warm, brace yourself against it, crack it open a little and be ready to slam it shut on the first hint of smoke. If a window is stuck during a fire, teach your family to escape by breaking the glass with a chair or dresser drawer and put blanket over the sill before climbing out to prevent cuts. Show children how to get out of a window in case of a fire. This instruction should also include evacuation from a second story window. Keep shoes by your bed so you don't have to walk on broken glass. Teach the whole family how to use the fire extinguisher.

During a Fire

During a fire, stay calm and rouse all house occupants immediately. Follow your plan and get out of the building immediately and shut the door behind you. Don't waste time on dressing or collecting valuables. Get the family together and keep together. Get out first and then call the fire department from a neighbor's phone. Many people die each year doing the opposite; that is, trying to phone the fire department from inside a burning building. Don't go back into your burning home; instead, account for all your children. Sometimes the children try to go back inside to get their favorite items or family pets.

If your clothing catches fire, don't run; rather, cross your arms touching your shoulders, drop and roll back and forth to smother the flames. It is usually best to run cold water over minor burns to cool them down. Don't clean a serious burn and never put grease, butter or ointments on a burn. Always call for medical help. If grease catches fire, don't put water on it. Use a proper fire extinguisher on it or smother it. If your cooking pan catches fire, cover it and turn the burner off. Never try carrying it off the stove.

If you are in a hotel fire, first call the fire department (give your hotel's street address) before you call the management. During a fire, don't use the elevator; use the corridor or stairs instead and take your room key with you. If the corridor is impassable, stuff clothes under the door, break the window and yell for help. Don't break the window if you are safe and there is smoke outside. This action would just let smoke in. Even if it is clear outside, you may need the protection of the window if smoke becomes a problem latter on. Watch out for flying glass if you choose to break the window. If you must walk through a fire, soak a towel and cover yourself. Cover your face, nose and mouth, but leave a little slot to see through. Breath through the wet towel. Almost all the fire department's largest ladders reach only to the seventh floor, so try to stay in a room at or below this floor. If help cannot reach you, close off all vents including the air conditioning and heating. Fill the tub in the bathroom with water and soak towels or sheets and seal all vents and cracks around the door. Try to use the phone to call the desk clerk and fire department. Shine your flashlight out the window. Leave your room only when you have access to a designated place of refuge or stairwell. Wait until a fireman arrives, for your chances of survival are much greater if you stay calm and stay put. You might fill the tub with water and get in. Firemen are trained to do a room by room search, so listen for their knocking.

Prevention

Appliances have additional fire risks. Keep your heater free from dust. Use a spark arrester on your chimney. Only use alarm clocks that have an automatic shut off. Never put tennis shoes, foam underwear or rubber products into a dryer. Make sure vaporizers don't run out of water. Never leave a hot iron unattended. Never place aerosol containers on stoves, heaters or in direct sunlight. In the kitchen, keep the range and oven clean. Clean behind major appliances at least once a year. Make sure you clean under the refrigerator at least once a year. Turn stove burners off or to low when you leave the kitchen. Pull out electric cords by their plug not their cord. If toast is stuck in the toaster, unplug the toaster before you try to get the toast out. Don't leave coffee makers on since they over heat when the water evaporates. Be sure all electrical appliances are UL-listed.

Did you know extension cords can cause fires? You should never plug too many electrical appliances into an extension cord (usually no more than one) and don't use worn out or damaged extension cords. Many people do not know electricity can jump from a wire and burn or kill you. Stay away from power poles; you don't even have to touch the wires for the current to jump to you.

Don't smoke when lying down, when your judgement is impaired by drugs or alcohol or when you are fatigued. It is important not to smoke around upholstered furniture when you are sleepy. Be aware that smoke from upholstered furniture can kill you before the heat wakes you up. Use ashtrays that will keep a burning cigarette inside and prevents the cigarette from falling out.

Special precautions should be used for fire hazards (i.e., brush, space heaters, lawn mowers, charcoal fluid and all flammable liquids). Check to see if you have any fire resistant walls and use them to your advantage by storing your flammable's on the other side of them if possible. Eliminate or beware of hazards like a messy work shop with exposed flammables. Attic fans should have an auto shut-off. To prevent Christmas trees from drying out, they should be watered and you should turn off the tree's lights before you go to bed. Remember gasoline fumes can burn or explode; hence, never clean anything with gasoline and never use or keep gasoline indoors. Make sure your fireplaces are completely screened. Only use commercially available starter fluids to start open fires or barbecues. You should keep doors shut while sleeping. Never go to bed expecting a fire, even in a fireplace, to burn itself out. Don't leave candles burning unattended. You should have corrective landscaping consisting of at least a 100' fire break between your home and brush.

Keep small children away from hot liquids. Even hot water from the faucet can cause serious burns. Set your hot water thermostat to low or 120 degrees if you have children in the house. It is common sense not to leave matches or lighters where children can reach them.

During a Earth Quake, Flood or Natural Disaster

We cannot prevent natural disasters but there is much we can do when we encounter one. When a natural disaster is happening stay indoors, get under a table or desk or stand in a doorway. After a disaster check for injuries. Don't use your telephone unless you are injured. Wear shoes. Turn on a battery powered radio for information. Check gas, water, and electrical lines. Switch off your power and don't go near downed electrical lines. Check your sewage lines to see if they are intact. Check your food and water supplies. Use water in your toilet tank, melted ice cubes or water stored in your hot water heater for emergency drinking water. Check your building for cracks. Check the chimney for cracks that could be a fire hazard and check closets and cupboards (open these cautiously). Stay out of damaged buildings. Do not make up or spread rumors. Don't go sight seeing and be prepared for additional occurrences. Depend on the radio and television for information and advice, instead of your phone. You can find the location of your nearest civil defense shelter and their evacuation plans by looking under civil defense in the phone book. I suggest you do that now.

Tornado Preparedness

About the only place safe from a tornado is entirely away from it. The best places to be, if you cannot avoid a tornado, is to be in a storm cellar, steel reinforced concrete building or bomb shelter. It is best to avoid open areas and upper floors of large buildings.

Cellars in brick homes are not safe, nor are vehicles a safe refuge. If you are in a car, drive at a 90 degree angle away from the tornado. Being caught in a city is the worse place to be. Get to the lower level of a steel reinforced concrete building if you can. Stand near a wall but away from the windows. If your caught outside, try to find a ravine or depression to lie in. If you are in an industrial plant and have the time, you should shut off all the electrical circuits and fuel lines.

If you get caught at home, you should get to the center of the building and stay in a small room or closet. You can also crawl under heavy furniture or stay under the stairwell. Stay away from chimneys. If you have time, open some windows and doors on the side away from the approaching tornado in order to help equalize the pressure on both sides of the house. If you live in a mobile home, go to a shelter.

Children at school should be taught to rearrange their desk so that the solid areas of the desk face the windows. Children should stay underneath their desk until it is safe. They may also go to a hallway and sit with their backs against the wall. They should evacuate flimsy buildings, and go to a ravine and lie flat. It is best to be protected by shelter, for heavy rain and hail usually accompanies hurricanes.

For tornado preparedness, you should have a battery radio, gloves and heavy shoes handy. Watch for fallen wires and don't smoke or light matches, for there might be broken gas pipes.

Things to Store in Case of a Natural Disaster

Perhaps you think a natural disaster isn't likely to happen tomorrow, the day after tomorrow, this year or in a couple of years; in fact, it can happen. They have happened in the past and are likely to occur again. Can you imagine what a major disaster would be like? Are you prepared? Storing supplies in your house will help reduce your hardship during a disaster. Taking the supplies with you to a shelter will not only help but also help you enroute. In the back section of this book there is a list of supplies that will help you cope with emergencies. For your own good, take the list seriously.

If you live in a flood zone, Securityland can furnish you with a free one page flier entitled "TIPS: Keeping the Inside of Your Home Dry in a Flood." All you have to do is mail us a self addressed stamped envelope to the following address:

Securityland
"Home Flood Tips"
P.O. Box 2079
Los Gatos, Ca. 95031-2079

Chapter 8

INSURANCE

With your insurance agent you should discuss what your policy covers. Coverage can differ greatly. Be leery of insurance that looks like a great bargain; can have many hidden provisions and limitations. If you do not know the meaning of something on your policy, ask your agent to explain it until you do know.

Insurance and its Various Types.

Insurance is a means of protection where a group of people pool their resources to protect themselves against an infrequent disaster. Each person pays a small sum to protect against their being the unlikely person to suffer a tragic loss. Here is a list of the most common types of policies.

"Normal Property Coverage" usually includes your house and attached garage, additions and other buildings on the property (except if used for commercial or rental property). It also covers personal property even when away from home. It can also include living expense money if your residence is rendered uninhabitable.

"Basic Form Coverage" insures against fire, explosion, lightning, riots, windstorms, hail, smoke, runaway vehicles, aircraft crashes, theft, vandalism, malicious mischief, house glass breakage and loss due to looting.

"Broad Coverage" has everything the "Basic Form" provides but also covers damages caused by weight of ice, sleet and snow, falling objects, collapse of buildings, steam and water systems, a few types of electrical equipment failures, plumbing freezing and heating and air conditioning failures. Content broad form insures only the contents of the house.

"Comprehensive" or "All Risk" insurance includes the benefits of everything already mentioned plus almost all accidents except landslides, war, nuclear radiation and surface water intrusion (floods, waves, tidal water, sewer backup and seepage).

A typical homeowner's policy protects you against loss due to burglary for up to 50% of the policy amount. For example, if your home is insured for $50,000, you would have $25,000 in protection on the contents of your home. If you rent, the full amount of what is stated on the policy is your coverage. There is 4 in 1 insurance which is $25,000 personal liability, fire, theft and vandalism. A home owner's policy can combine fire, liability, burglary, some hazards, 80% co-insurance, theft and vandalism. There are many policies (e.g., 1 to 1, 1 to 16 and 18 risk type). Talk it over with your agent. Don't be fooled; shop for the best price for the coverage and service you need.

Most insurance policies work from 1-15 year depreciation schedules. For example, you may receive, as an insurance settlement, for the loss of a two year old TV half of what you paid for it new. There are also coverage limits that an insurance company will pay, (e.g., boat coverage limited to $1000; jewelry, furs $500-1000; firearms $200; check forgery $500-1000). There are varying rates for credit cards and named perils. All Risk insurance is replacement coverage and is usually a good value if it only cost 15% more than a regular policy.

Personal liability insurance protects you against damage claims by another person. It pays medical bills (usually up to $500), property damage claims (usually up to $250) and personal liability claims (usually good up to the amount of the policy). Personal liability coverage would pay your court defense expenses in case of a lawsuit. Family liability protection is good for injuries and property damage that occurs as a result of actions of your immediate family (e.g., a toy on the floor causing a fall of a neighbor).

The price you pay for insurance depends greatly on where you live, your home's condition, the coverage you want and the company you insure through. The only way to know for certain is to talk with several insurance agents.

Things to Check

1) Government insurance is available in most states. If you live in a high crime area this type of coverage might be best. You also cannot be turned down or be declined after filing a claim.

2) Coinsurance clauses in some policies say you must carry 80% of replacement value to guarantee full payment.

3) Supplementary protection and riders might be needed to cover theft that occurs when you are away from your home and for swimming pools.

4) Boat owners usually have a separate policy for their boats.

5) Floaters can be purchased for specific items above the ceiling; for example, an antique watch.

6) Flood and earthquake insurance is usually not covered without specifically asking for it. Think about getting some if it could happen in the area you live in.

7) Ask your agent about theft from your vehicle. See if you are covered for both what is attached to the car (radio) and contents (golf clubs).

8) Ask about risks not covered and how much it will cost to cover them. For instance, personal property, alarm system discounts, home owners, co-insurer clauses, riders, and what type of family liability coverage you have are some other types of coverage you should evaluate when selecting your coverage.

Recommendations:

1) Shop for coverage because rates vary. Sometimes you can get twice the coverage for the same price.

2) Insure to the amount of losses or accidents that would be a financial hardship for you to replace.

3) Get a policy with a deductible and large coverage. It is better to be covered for a total loss than save money by covering only small losses. Coverage of $25,000 can be raised to $50,000 for usually a small fee.

4) Get at least 85% replacement cost. There are policies for fire and burglar insurance that pay full replacement cost. Inflation and rising building cost can leave you under insured. Many policies have automatic inflation values built in. Review your insurance coverage every two years.

Chapter 9

CRIMES AGAINST CONSUMERS, CHILDREN, THE ELDERLY AND HOW TO REPORT CRIME

General

If a deal sounds like it is too good to be true, it probably is. Be skeptical of free or last chance offers. Do not rush into something involving your money or property. Never turn over large sums of money to a stranger for any reason no matter how good the deal looks. Don't buy large items with cash. Comparison shop for the best service or price and get second opinions before you invest. Talk to your bank officer if you are going to make a substantial withdrawal for a purchase or business deal. He or she can help you do it safely and wisely. Do not hesitate to check on the credentials of a salesman or public official. Don't pay for goods or services in advance and get a receipt for all transactions. Don't sign a contract until you have read it carefully and understand it or have a lawyer review it for you. Cancelling a contract is usually a good idea if you have second thoughts. Call the consumer protection agency, law enforcement officials, District Attorney or State Attorney if you have second thoughts. Report all suspicious offers to the police immediately. Go to the police if a swindler gets you. Report it before the swindler has a chance to leave town and victimize again. Delay could help them get away. If you are duped and you don't report it to the police, they are free to cheat again and you have no chance of getting your money back. Con artists count on their victim's reluctance to report the crime. I feel if you do not report the crime, you have been duped twice as bad.

Some Things to Watch For

Be cautious of a "great deal" (e.g., renting a house well below current rates). The advertisers might not own it and they rent it out to several people before skipping town. Watch out for bank examiners asking for your help. Real bank examiners don't use civilians but criminals do. After the death of a family member, criminals may try to deliver things never ordered by the deceased. People who say you need repairs when you didn't call them could be a fraud. They may say they were driving by and saw you have a bad roof (call your own repair people). Don't be afraid to say no to high-pressure sales tactics. Criminals many times will go up to a house and say they are undercover police and want you to check your valuables. Then they either rob you or come back for your valuables when your are not home.

Never invest in get rich fast, government land, oil and gas, miracle car or uncut crystal schemes. Don't believe claims of excessive weight loss, secret cures, miracle drugs or regained youth potions. A typical deal is like the ad I have recently seen "Hit pay dirt! Gold to share with friends! No dealers. Mail in just $99 for a 1 gram 24kt. gold ingot before they are all gone." Yes, they will send you a 24kt. 1 gram gold ingot but 1 gram is equal to 1.543 grains and there are 24 grains to an oz. of gold. This is about $3,000 per oz.; therefore, it is not a very good investment.

Crime and Children

Each year more than an estimated 100,000 children suffer some type of sexual abuse. Never assume your child will never be abducted. Teaching children about this crime is very important for their safety. Realize children are more vulnerable. Children often make stories up but rarely lie about being a victim of a sexual crime. In over one third of sexual abuse cases involving children, the offender is known by the child; so watch for signs of over affection by family friends and relatives. Realize a pedophile is usually an adult male with a sexual preference for children. They prey on runaways or children from unhappy homes. They shower your child with affection whenever the sexual activity takes place so that the child not only does not complain but also many times fails to report the crime. No one should care more about your children than you. Always show your child affection. Watch your child's behavior for signs of abuse. If they are more withdrawn or show signs of anxiety, talk to them. If a sexual crime happens against your child, have them checked for venereal disease and pregnancy.

Things You Can Do

As a parent there are many precautions you can take to prevent crimes and accidents from happening to your child. Children are greatly influenced by their parent's behavior. Children learn their security habits by mimicking their parents. If you always lock your door, so will they. You should make sure the child's name is not readable on clothing, caps, bikes, wagons, etc.. Mark it on the inside where it cannot readily be seen because children respond to strangers who know their name. You should also return early the first few times when using a new baby sitter. Tell the sitter not to let your child leave with anyone but you. Have strict procedures about who will pick up your child from school. Be consistent in your procedures and time that you pick up your child after school. Have your child's school establish a call back program to phone anyone who is absent. Have your child's picture taken yearly; for pre-schoolers, take pictures four times a year.

Keep all records of dental, fingerprints, footprints, doctor information, birthmarks, and their birth certificate. Get your child a passport because, once issued, it draws suspicion when anyone reapplies using your child's name (contact: Passport Services, Bureau of Consumer Affairs, Washington, DC.).

You should stay in touch with your ex-spouse and his/her friends and relatives. Pay attention to any threats of child abduction. If an ex-spouse has an attitude change, be cautious and be aware how these changes may affect your ex-spouse's behavior. Be involved in your child's activities.

Know where your child is at all times. Know the names of your child's friends, their phone numbers and addresses as well as who is responsible for supervising them. Know where your child is all the time. Point out places in the neighborhood for them to avoid. Don't let children go door to door canvassing unless accompanied by an adult. Accompany children in rest rooms or have a friend do it even on airplanes.

It is common sense not to leave matches or lighters where children can reach them. In fact, it is best not to leave small children alone at all. Teach children stealing is wrong. Keep small children away from hot liquids. Even hot water from the faucet can cause serious burns. Watch them when they are in the tub. When you cook or boil water on the stove, use the back burners so your children cannot reach the handles.

Do not store chemical poisons in locations readily accessible to children. Do not raise indoor or outdoor plants that are poisonous. Store a one ounce bottle of syrup of ipecac for each child or grandchild. This medicine, which induces vomiting, is nonprescription and can be obtained at any drugstore. Never use this drug unless directed to do so by the poison center. Always call the poison center first when you suspect a poisoning.

Don't leave a child alone in car, mall, shopping carts, toy section at a store or anywhere else. You should not leave small children alone even for a minute in a car, at home or any public area for a minute is all it takes for them to disappear.

If you lose your child, treat it like all other crimes by staying calm and thinking. Think about what the child was wearing, where he was, when he was last seen, who was near him and what was he doing. Report to the police if you see a kidnapping or know of or suspect a kidnapping. If your child is missing, look for clues at home in the child's room. Check the neighborhood, school, youth associations and clubs. Call all your child's friends and your friends, relatives, peers and distant relatives. Check out all urban areas, even locked ones like roofs, basements and garages. Check any nearby transit areas. Publish the disappearance with fliers and employ a private investigator. There is an organization that helps locate missing children. It is called The Missing Children's Network (800) 235-3200 and they offer many other helpful services.

Checklist for Child Safety

Teach your children early about the facts about abduction. Don't frighten your child. Just say it is another coping skill. There are many good books at public libraries that may help you with this.

Children are naturally friendly with strangers. Discourage this! Don't let strangers touch your children. Friends are guests in your house and are not delivery people and salesmen. They should learn police are friends and that police officers not wearing uniforms and driving police cars will not pick them up. Friends are not people only known by sight. Strangers are everyone you don't know well. Teach them to never accept rides from strangers. Teach them to never say, "yes" to strangers offering gifts; instead say, "No thanks." Have children refuse unnecessary requests from strangers. Teach them to politely say "You can manage without me." If a stranger in a car asks directions, don't go near the car. Tell your child that adults usually don't ask directions from children. Don't go near a stranger's car even if signaled. Avoid cars parked with the motor going. Teach children how to react to a stranger who does things he should not do. Avoid strangers who are hanging around; especially avoid an adult who wants to play games with them and their friends.

Don't go with men to their living quarters. Keep windows and doors locked. Don't open the door to strangers. If a delivery person is at the door and they are home alone, instruct them not to open the door. Instead, tell the delivery person through a window to leave the package on the doorstep. Tell them never give their name or address to a stranger.

If followed, don't hide behind bushes or isolated areas. Go to a place where there are people. If a threatening stranger approaches them on street, step back, turn, and run in the opposite direction. Tell children to make a scene when threatened by an adult; kick, attempt to break loose and yell "help"; do not just scream. Tell them not to fight if they see that the attacker has a weapon.

Have a code word that only you and your child knows and teach your child not to go with anyone not knowing the code word. Change it if you ever use it. If someone claims you sent them to pick up your child from school, tell your child to go back into the school for help. Children should walk the same path home from school every day. Explain that abductors may tell lies (e.g., "your mother and father are dead," or "they do not love or want you anymore").

If a stranger bothers them, they should run to a policeman, a neighbor, teacher, parent or some adult they know. Children should never fight a bully or a group of bullies. Have children tell adults instead. If something is strange or unusual, they should tell others. No one should take pictures of them when they are naked and any stranger taking pictures of them is unusual.

Children should inform parents of their whereabouts. Older children should use the phone and call if they have a change in plans. Tell them where to go if they get in trouble.

Teach your children how to recognize sexual abuse and to recognize what is wrong. Tell them no one has the right to touch any part of their body and if someone tries tell you. If another adult tells them to keep a secret, they should tell you.

If they are a victim of a crime, they should be taught it's not their fault and that they didn't do anything wrong. Tell them they are not bad when someone harms them. They should report problems immediately to you. Always have open communications with your child so they will confide in you during trouble.

Have them learn their full name, address, and phone number including your area code number. They should learn your full name and your office phone number. Teach them how to make long distance phone calls through direct dialing, operator assistance calls and 911 calls. Know emergency numbers like O, 911 or whatever number is used in your area. Teach them when it is appropriate to use them. Teach them to dial the operator if kidnapped and tell the operator the number on the phone they are using. They should leave the phone off the hook for help to arrive. They should always carry a quarter or two dimes for emergency phone money.

Teach your children never to say they are alone over the phone or at the door. Instead, have them say that their parent is napping. They should be taught to carry money in their pockets until needed and not show it to others.

Children should come home before dark and never be out at night. Tell your child not to go into anyone's home without your permission. Tell them never volunteer future family vacation plans. Also tell them not to play in deserted areas (like the woods or in vacant buildings) don't take short cuts through alleys, don't loiter in public rest rooms or elevators and especially in parking lots. Children should walk and bike with friends (use the buddy system).

When children are alone they should avoid movie theatres, arcade game parlors, libraries, parks and swimming pools.

Teach children to remain within eyesight when you are shopping. If children become separated from you while shopping, they should know to go immediately to the nearest sales counter and ask for help and not to walk around looking for you. Children should know about chemical poisonings and warn them not to eat or drink anything without your approval.

Teach them to ask anyone transporting them not to leave them in the car alone. If it cannot be avoided, they should sit in front near the horn, lock all the doors and roll up the windows leaving just a finger space. If a stranger tries to enter, instruct children to blow the horn until help arrives.

Role play with your child. Bad guys say "Do you want some candy?", "Mommy needs you. Come with me.", "If you come with me I will give you a present.", "I have lost my puppy. Will you help me find him?" and "I am a policeman, and I am here to take you to your parents."

Tell the children that, if anything happens, you will look for them no matter how long it takes.

Crime and the Elderly

The elderly should read this book entirely. All the usual crime prevention measures apply to them. Elderly are victims of crime less often but it is usually more severe. Planning ahead and being aware of their limitations can help avoid possible problems. Safety should be the main concern. If their memory is faulty, they should always carry identification or mark their clothing with their address and the phone numbers. Check with the local senior citizen center or church for escort and protection services. Traveling in groups is highly recommended for the senior citizen. Talk to elderly people daily on the phone and visit them regularly. Visit their neighbors also. Have social security checks and retirement moneys automatically deposited in the bank. Ask the bank about the Treasury Departments Direct Deposit Program for this service. Help them develop a system of counting money if it is a problem. They should keep their shopping money separate from their transportation money. They especially should only carry the money they need and sew special pockets in their clothes to carry their valuables and money. Always use pockets rather than purses for valuables. Be aware of anyone offering medical services at the door or over the phone at a discount. It probably is a fraud. Elderly people should also report all crimes even if their memory is not perfect; it can let police know there is a problem.

Reporting Crime

Always get involved and report all crimes against yourself and others. If you see a crime but are afraid to report it, you can report it anonymously. Call the 800 number operator at (800) 555-1212 and ask for the nearest We Tip number. If there isn't one call your local police.

If you return home and see it has been broken into, go to a neighbor and call the police. Don't go inside until the police arrive. After a burglary, do not touch anything until police get there. If victimized, the very first thing to do is to call the police even before you call a friend or relative. Try to remember the criminal's car license plate and write it down. Try to get the first numbers or letters if you cannot get the whole thing. Reporting crime immediately increases the chance of catching criminals. A five minute delay reduces the chance of catching a criminal 66%. There is a chance that the criminal is still around and can be caught.

After you phone the police, try to remember the criminal's physical appearance (see form included). Write down only things you are sure of; features are more important than just clothing. Look for the sex, race, age, jewelry, weapons, teeth missing, glasses, birthmarks, scars, handedness, accent, tatoo, height, weight, hair color, facial hair, complexion, eye color, nationality, clothing, shoes, voice character and unusual body gestures of the criminal. Remember their vehicle's description make, year, body style (2 door), color, dents and its license number. Write down everything you can remember about the incident. How many suspects were there? What did they do? What did they say? What did they take? Which way did they go? Were there any other witnesses? If there were witnesses, get their names, phone numbers, and addresses. Write down anything else you feel is important.

After a burglary, talk to the police, your neighbors, insurance company and your credit card companies. Get officers names and ask them whom you should contact if you should remember or discover information after filing the report. See what is missing and clean up only after the police leave. Take your time in filling out the insurance forms and don't let the insurance agent rush you. It is in your best interest to make sure everything you lost you get reimbursed for. You should decide who will clean after a fire instead of your insurance agent. Talk to an accountant; what wasn't covered might be tax deductible. Don't expect to get your property back; out of $400 million worth of goods taken yearly only about 5% is recovered. It still cannot hurt to follow up on the police report with a call now and then for they might recover some property. Make any security corrections you can to insure it will not happen again.

Many burglars return to the same house five months after a burglary to rob it again for the new valuables from the insurance company's reimbursement. They know the house layout and its security defenses. You may experience emotional shock after a break in: don't be to shy to get professional psychological help if needed.

Chapter 10

CRIME STORY

Cops seek rapists

An 18-year-old San Jose resident was kidnapped and raped several times by two men when her car stalled on Blossom Hill Road near Vasona Park at 5:45 p.m. Wednesday, March 19, according to police.

The young woman was going to visit a friend in Los Gatos when her 1969 green Dodge Dart stalled. Two men in their 20s driving a red pickup parked behind her car. One man reached his arm through the partially opened window on the driver's side, unlocked the door, forced himself into the car and held the victim's head down. The second man got the stalled vehicle to start.

They drove to the Summit Road exit off Highway 17, where they raped the victim several times, said Detective Jerry Meyer. The men then drove the woman back to where their red American full-size pickup was parked around 9 p.m. According to Meyer, the suspect's truck had right front end damage.

The two white men are described as follows: one is 25-30 years old, about 5 feet 8 inches tall, with a scar on his left cheek, straight shoulder-length blonde hair and a muscular build. The other has been described as 20-25 years old, about 6 feet tall, with dark brown wavy hair, two days' growth beard and a tattoo on the back side of his left hand. He has a thin build.

Anyone with any information on either of the suspects or who may have seen the red truck or Dodge Dart should call Meyer at 354-6825

Crime never has a happy ending but this could have been much worse. This is not a story but a real crime. This despicable kidnapping and rape occurred about 1/4 mile from my house and, as far as I know, the suspects are still at large. Do not think this is an unusual isolated incident. It is one of the many crimes that happen every day. The 1986 statistics show one out of every three women will be raped in their lifetime.

With the steps covered in this book you can possibly avoid this from happening to you.

Review-

1 Travel with a friend whenever possible.

2 Keep your car in good repair and do not let the gas tank go below the 1/4 full mark.

3 Avoid driving on desolate routes when there are other roads.

4 Keep all the doors locked and windows rolled up when traveling alone.

5 If your car breaks down, just roll the window down an inch to tell people to get help for you.

6 Carry a noise alarm device when in dangerous situations.

7 Think about resisting: fight, talk to them, act strange, etc.

8 Report to police.

9 Contact rape crisis center.

10 Seek psychological help.

How Crime Has Touched Me

This is my partial list. Crimes like this happen every day. See if you can remember the prevention techniques.

Grade School

1) A close friend of mine was raped in a park when he was 10 years old.

2) A peeping tom was seen outside my sister's window.

3) My brother's bike was stolen from a convenience store.

4) Thieves tried to steal our bikes from our porch.

5) An older kid recruited me to help in burglaries. He was caught and charged with stealing $80,000 worth from homes. He was sent to a detention center.

6) My neighbor died from a exploded flammable furniture polish storage container.

Junior High

1) I had a switch blade pulled on me by a fellow student at school over an argument about a basketball.

2) My mother was molested on a public bus.

3) I saw a holdup at a drug store.

High School

1) I saw the police removal of a murder victim's body at a park.

2) A classmate was electrocuted by a power line.

3) I sold guns at a sporting goods store. I saw many, victims who decided to arm themselves (people from 21 to 69). I sold guns to many police officers and heard many stories. I sold guns to average looking people who turned out to be criminals.

4) Our car was stolen but was returned by the police the same day.

College

1) A friend was brutally raped and she never returned to school.

2) My girlfriend was approached by a married professor in order to get her to sleep with him.

3) On my summer job I stayed at a dive motel. I saw hatchet fights, drugs and beatings. I became a guardian to the children being brought up in this hideous environment.

4) My roommate had his wallet stolen with all his credit cards. (He was in a nearby pool).

5) Someone stole fishing equipment from my car. They broke the window to get it.

This Year

1) Mother of my little brother (I am in the big brother program) had her purse stolen.

2) Arson fires were set. One at my old high school and another burned for three days and took many homes.

Make your own true to life list right now and check to see if there was anything you would have done differently.

Chapter 11

CONCLUSION

We should not have to live in fear! There is only a small minority of criminals and if we act as a group with the police we can easily combat them and live fuller lives. This book is only good if you use the suggestions. Remember, the one time you ignore the precautions might be the one time some act of crime happens to you. Many of the recommendations may sound silly but crime is not a joke. It is better to act a little weird than to suffer as a victim. Security measures after the fact aren't as good. If you must change your habits for protection, change your habits. Start now; there is no other way. Avoiding crime and escaping with the least amount of harm to yourself is well worth the inconvenience of learning the simple avoidance measures mentioned in this book.

Things You Can Do

Continue your effort to avoid crime (you already took the first step by reading this book). Review this book periodically; pick a date, a yearly holiday for instance, and mark it on your calender (review it before your celebration). Don't buy stolen goods in order to dry up this market. If something is a great price, be suspicious. I know one person who bought a T.V. from a fellow out of a truck but, when he got it home, the container contained a broken, outdated, used T.V. (what a deal for $250). Possession of stolen property is a crime. You should ban together with neighbors and look out for each other's interests regardless of personal differences in order to stamp out criminal activity. In order to get crime and violence off television, write letters to the stations and your political officials. Become informed on your local judges performance to vote for the ones who share your views, join the civilian police and report crime.

In the Community

Realize that crime is a community problem. Ban together with neighbors, police, other citizens and the community. Together we can stop crime. Have police involvement. Call your local police or sheriff's office and ask about their crime prevention programs.

State and Federal Action

Remember, we elect the officials. We vote for state personnel who handle the state courts, drug programs and corrections. We vote for federal personnel who hire the personnel for the federal courts who in turn handle corrections, terrorist, organized crime, white collar crime and international crime. The justice system including courts, police, and corrections run up a yearly bill of $34 billion. See to it that they use your money wisely.

Causes of Crime

No one is sure what is the cause of crime. Everyone has their own opinion (i.e., economic inequalities, gun ownership, immigration, urbanization, loose laws, police politics, ineffectual law enforcement, media glorifying criminals, over population, disintegrating nuclear families, over crowded schools, lenient courts, and even prisons. There is no consensus; however, everyone agrees crime has no simple cure. You cannot raise education, increase employment, raise the living standard, create jobs, stop drug smuggling, reestablish the nuclear family, and unite the general populous to fight crime all in a day. Solutions take time.

More Opinions

First time offenders are unlikely first time offenders but instead were apprehended for the first time. Probation is too light for first offenses. I would like to see more street crime units, sting operations, decoys, home alert programs, truancy abatement programs, and citizen watch programs as police priorities. I think judges should have the right to pre-screen prior relationships (criminal and victim) and witnesses for intoxication at the time of incident. Also, I would like judges to interview and instruct on the consequences of filing, as well as presumptive sentencing. This would throw out many cases before they waste time and money.

There should be a minimum sentence for all violent crimes. There should be a minimum one year sentence for carrying a gun without a permit in order to get guns off the street. For nonviolent first time offenders, judges should not give long prison terms; they should give only a slight taste of it. Our prisons should keep violent criminals locked up and not let them out early because of over crowding. Judges should be able to look at juvenile arrest records of five or more arrests. This would flag the career criminals early before they do serious damage. Judges should be able to rule on arrest on probable cause of findings. There should be little or no rehabilitation for prisoners; it is very costly and doesn't greatly reduce the number of repeat offenders. It is great in theory but it just doesn't work. I also feel there are many people in prison who are nonviolent criminals who also could be working in the community, paying back the victim or doing community service. This would also benefit society by cutting down on the high cost of running the prisons. I am not an expert but I believe the statistics I have examined. The remedies I have mentioned warrant a try.

Other Topics

Crime is over a two hundred year old problem for the U.S. and it will not be stopped over night; however, it can be greatly reduced. This book concentrates on prevention. There are many subjects that I have omitted because I wanted to make sure that this book is as brief as possible. Subjects not discussed or only briefly mentioned that should be examined include: white collar crime, schemes, scams, bunko artists, business crime, bad check writers or domestic violence, terrorists, and international crime. Other subjects of interest include: psychological effects on victims, penology, law enforcement coping, professional criminals, ex-offenders re-entry into society, crime deterrent architecture, citizen watch programs, the judicial system, self defense, juvenile delinquency, and rural crime. Cases of criminal behavior are not discussed in this book. See the selected readings section of this book and visit your local public library to become more aware of crime and more secure.

Appendix A

SELECTED READINGS

There are many good books at bookstores and libraries on subjects including rape, child protection, home alarms, locks, dog training, bunko schemes, natural disasters, fire prevention, insurance and criminals that can make you more informed on these subjects. The following selected readings have valuable information for the crime wary.

<u>Selected Books</u>

Forest, Jr & Martin L. & Estrella, Manual M. "Family Guide to Crime Prevention." Beaufort Bks. N.Y.,: (ISBN 0-8253-0036-3) 1982. 253p.

Lewin, Tom. "Security: Everything You Need to Know About Household Alarm Systems." Park Lane Ent.,: (ISBN 0-96093620-3), 1982. 99p.

Morton, Bard & Sangrey, Dawn. "Crime Victim's Book 2ed." Brunner-Mazel,: (ISBN 0-87630-415-3), 1986. 272p.

Newman, Oscar. "Defensible Space: Crime Prevention Through Urban Design." Macmillian,: (ISBN 0-02-000750-7), 1973. 264p.

O'Block, Robert. "Security & Crime Prevention." Butterworth,: (ISBN 0-409-95138-2), 1981. 452p.

Roper, A. "Complete Book of Locks & Locksmithing." TAB,: (ISBN 0-8306-1530-7), 1983. 352p.

Sorrento, Anthony. "Organizing Against Crime: Redeveloping the Neighborhood." Human Sc,: (ISBN 0-87705-301-4), 1977. 272p.

Time-Life Books. "Home Security." Time-Life (Home Repair & Improvement Ser.),: (ISBN 0-8094-24185) 1979.

Appendix B

VACATIONER'S CHECKLIST

(Special Precautions for Special Occasions)

Read this list whenever you are about to take a vacation. Tear it out of the book and take it with you when you leave if it will help. You cannot remember it all.

Burglars prefer looting a house that is vacant; therefore, making your house appear occupied at all times greatly reduces the chance of being burglarized. Before your next vacation, be sure to use this checklist.

Check the following items off as you go down the list.

Before You Leave

1. Have a trusted neighbor or pay a neighbor's child to daily pick up all items that are delivered at your door while you are away. If a friend is not available, have deliveries temporarily canceled.

2. Make the yard look like it is used. Make sure someone cuts and waters the lawn and shovels the snow or walks up to your front steps through the fresh snow.

3. Arrange garbage to be put out and picked up as usual.

4. Have your pets fed and watered at your home instead of taking them to a kennel; pets provide additional protection while you are away and make your residence appear occupied.

5. Set thermostats according to normal living conditions that fit the current weather or put your air conditioner on the fan setting and plug it into a timer.

6. Hanging laundry out, when going on a short vacation, is a excellent deception for making your house look occupied.

7. Have your second car in the driveway periodically moved.

8. Store valuables in a safe deposit box or at a relative's.

9. Place your ladder and valuable yard items, (e.g., tools, lawn furniture, etc.) in the garage.

10. Lock up all gates especially at night.

11. Avoid pre-vacation publicity about your trip.

12. Leave instructions on how to turn off the alarm with a neighbor; don't forget to give them the key.

13. Inform law enforcement about your trip and leave them the names, phone numbers, and addresses of anyone who should have a key including your neighbors.

14. Don't pack the car the night before the trip.

15. Have your car checked out (now is the time to fix that funny noise that you hear. Read your owner's manual and follow its service instructions. Check on tune up, brakes, brake fluid, windshield wipers, lights, horn, steering, hoses, belts and battery. Have the gas tank level above the 1/2 full mark. Check your cars oil (keep an extra quart in car trunk), your tire's wear and inflation, radiator water and coolant, transmission fluid and exhaust pipe. Have a jack, tool kit, flares, blanket, matches, umbrella, spare tire, flashlight and maps in the car.

16. Use a reliable source to gather information about your trip. AAA or travel agencies can tell you what to sight-see and what you have to be cautious about.

17. Pay bills that will come due in your absence.

18. Buy and use traveler's checks. Record the amounts and numbers on a list and leave a copy of the list at home.

19. Remove all valuables of note from the house and place them with a relative.

20. Have credit card and traveler's checks and the company's phone numbers in case they are ever lost and take your insurance company's phone number and policy number with you.

21. Consider securing the services of a citizen watch patrol or house sitter.

22. Check to see if you have insurance coverage-both medical and property while on vacation.

23. Don't leave a pile of trash or leaves in the yard to tempt an arsonist.

24. Buy a travel lock for your hotel drawer.

25. Buy a travel alarm for your hotel door.

26. Get a small flashlight for your hotel room and a flood flashlight for your car that can be used as a self defense weapon also.

27. Put on suitcase stickers (e.g., outside stickers should be bright, inside stickers should have your last name and address).

28. Get a passport and immunizations if needed.

29. If going abroad, check with foreign embassy about their laws. If your are going to drive, get road maps at your local automobile association. Get an international drivers license if needed.

30. Don't use expensive luggage; thieves will steal it just for the value of the luggage alone as well as for the belief there are nice valuables inside.

31. Do not over pack suitcases. Don't put fragile, vital and important items like medicines in your luggage. Irreplaceable items like jewelry, money, negotiable papers and art should be put only in carry on luggage. Airlines are not responsible for these items. Airlines are usually only responsible to pay a maximum of $750 dollars for your lost luggage and items and the airlines depreciate the value of these valuables. (e.g., a $350 suitcase bought two years ago would only get about $125 in compensation). Lost luggage is the airlines number one complaint. Airlines typically will reimburse you half the cost of new clothes and if you turn in your old clothes when they are found, you will be reimbursed completely. You have to notify the airlines right away if you lose your luggage and you have 45 days to sign a formal report. If you have any complaints, address them to the Civil Aeronautics Board.

The Day You Leave

1. Pack your car the day you leave; do the packing inside your garage and place your valuables in the car trunk.

2. Lock and leave your second car in the driveway.

3. Lock your garage door.

4. Arrange drapes so police and neighbors can look inside.

5. If possible have a person raise and lower drapes routinely.

6. Turn down the volume control on the phone so it cannot be heard from outside. If you cannot reduce the volume, either disconnect your phone or have the phone company redirect the calls when you are going on an extended leave.

7. Set up radios and lights on electrical timers. Program timers to turn on a radio talk show in low volume and lights to come on at variable times.

8. Leave a light on at night; dark houses stick out to burglars.

9. Leave an itinerary with a neighbor and relative so you can be reached in an emergency.

10. Remove from your wallet all the papers and credit cards you won't need on the trip.

11. Check your alarm system.

12. Lock up all doors and windows.

13. Do a double check. Walk around the house; doors should be locked and the spare key should be picked up.

14. If taking a taxi to the airport with your luggage, and the taxi driver asks about your trip, say it's a short trip and that your house will be occupied in your absence.

15. Drop baggage at the terminal first. Leave it with friend or baggage handler. Dragging baggage around the parking lot makes you a target.

While on the Road

There are criminals who specialize in robbing vacationers because many people let their guard down while they are having a good time.

1. Use traveler's checks.

2. Have an escape plan in case of fire. Locate the fire exits as you go into hotels and motels.

3. Sleep with your hotel key and flashlight near your bed and take them with you during a fire.

4. If injured, take steps to protect your valuables before being admitted to the hospital. Give them to a friend or relative.

5. Bring all your luggage into your hotel room at night.

6. Determine taxi fares with the driver before you use this service.

7. Don't leave valuables like cash, credit cards, checks, jewelry or cameras in rooms since most states only require lodging to carry a maximum $250 liability insurance. Check valuables into the hotel vault. Store reserve money and credit cards there also.

8. Have wearable identification for children. Use a marker on the inside of their clothing.

9. Claim luggage as soon as possible at airports.

10. When traveling through an isolated area, tell a person where you will be and when they can expect your call.

11. Remember, there are many people with pass keys in a hotel.

12. Be especially cautious when in the shower.

13. When eating on the road, park where you can watch your vehicle.

14. Don't advertise you are a tourist (leave maps and travel brochures in your glove box).

15. Don't take maps into restaurants.

16. Keep travel brochures and maps in your glove box instead of on your car seats.

17. If you split up from your group, keep track of the time and return promptly.

18. If you need to abandon your car, keep everyone together.

19. Ask directions before you set out. Looking lost makes you a target.

20. Never open your door to strangers at home or on vacation.

21. Watch out for pickpockets in airports, hotel lobbies and crowded areas.

22. Don't carry a lot of cash and divide it among several pockets.

23. Carry cash only in small denominations.

24. Don't carry cash inside suitcases.

25. Hang the "do not disturb" sign on your door when you leave your room if the maid is finished.

26. Always lock your room's door even when inside.

27. Don't tell people in the lobby or at the pool your room number.

28. Don't invite people to your room; instead, meet them in the lobby.

29. Be leery of single, lonely people; they might not be either.

30. Stay away from the seedy parts of town.

31. If you see a crime report it. You want others to do the same for you.

32. Vacation time is a high crime time. Stay alert.

33. Only ask directions from a gas station attendant or a reliable source.

34. Use only well traveled roads even if there is a short cut.

35. Camp in only approved camp grounds.

36. Disembark from a ferry immediately after it docks.

37. Don't take tours unless given by resorts or public agencies.

38. Be leery of strangers asking about your vacation plans.

39. Watch out during parades and festivals.

40. Don't stray from tours.

41. When inside an R.V. camper, always keep the back door locked.

42. Watch your luggage in no man's land between the curb and the hotel lobby.

43. When you are attending a convention, remove your badge when you go out of the conference room.

44. When traveling abroad, know a few words in the native tongue in case of an emergency.

45. When you unpack, place belongings in the closet and dresser. Arrange it in a way so you will know if anything is missing.

46. Consider locking electrical appliances (e.g., electric shavers and hair blowers) in suitcases where they will be harder to steal.

47. Always keep suitcases locked even if they are empty so they cannot be used to carry out your property.

48. Report any suspicious acts in corridors. It is the best way to fight crime.

49. Don't leave valuables in a rented vehicle; likewise do not leave them in the trunk.

50. Always have someone stay with your valuables even if you want to swim at a beach.

51. When leaving your room, turn on a light and the radio or T.V. to give the impression someone is in the room. During the evening leave a light on.

52. When taking a taxi, don't volunteer that you are a stranger to the area to the taxi driver. This could tempt them into over charging you or taking you on a longer route to your destination.

53. If you are a woman traveler and are registering at a hotel, use only your first initial and last name.

When Returning Home

1. Phone ahead to your house watcher to report your arrival time.

2. Do a walk around before entering your residence and, if you see evidence of a forced entry, go to your neighbor and phone the police.

3. Check your smoke detectors to see if the battery is still working.

Appendix C

SECURITY & SAFETY CHECKLIST

This list helps determine areas of special attention but the list is not all inclusive. Each environment can differ. This list helps determine the weakest link a burglar might use. Most burglars aren't professionals. On the contrary, most break-ins are committed by juveniles who see an open window, a faulty lock or an open opportunity.

This walk around security check goes from the outside to the inside of the home and discusses general items. Many police departments will do a home security analysis of your home free of charge. They can help point out weak points in your security system. Evaluate your home security from a burglar's point of view; will your house make an easier or better target then other homes on your block.

Outside the Home

Doors - Use the following checklist when you inspect all outside doors.

Use solid core construction or steel doors because burglars can literally walk through a weak door. Strong door frames are as important as solid doors. If there is glass in the door within 40 inches of a lock, use a double cylinder and lock and unbreakable safety glass. These precautions make sure a burglar can't break out a panel or glass in your front door and unlock it.

If hinges are facing in, have a special screw or pin installed. Does your door have a peephole with a wide angle viewer? Secure your door if it has a breakable glass panel. Doors should have a strong cylinder dead bolt, 1" throw with a beveled cylinder guard and a striker plate with 3" screws. Doors without cylinder locks should have a heavy bolt. It is best to have double cylinder locks. Make sure all doors can be securely locked. Basement doors should have locks to isolate that part of the house from the rest of the residence. Secure all doors: basement, cellar, porch, balcony and sliding doors (charley bar). Make sure your locks and hardware are in good repair. Were your locks rekeyed when you moved in? If not, do you know everyone who has a key? Do your out-swinging hinge doors (reverse handed doors) have pinned or non-removable pins? Sliding doors should have auxiliary locks. Use a mail slot if possible.

Keep doors locked even when home. Chain latch your front and back doors.

Windows

Are all your windows equipped with auxiliary key locks? If windows just have thumb screw type latches, add a key lock. Secure or replace louvered windows. Window locks should be securely mounted. Keep windows locked when shut. Use locks that allow you to lock windows when they are partially open. In high hazard locations, use bars or ornamental grills. Put locks and curtains on garage windows. Use the same precautions and secure all basement and second story windows like all other windows. Make sure your windows are not removable. Secure all windows with lag bolts or have grills placed over them (even small windows). Remove all valuables seen through windows. Buy an alarm if you are worried. Make sure windows can't open easily by breaking a small glass panel. Lock all windows before you go out. Nail your window down on your air conditioner.

Secure all other openings: skylights should be locked and alarmed, roof openings locked, fire escapes should have extra locks preventing ingress.

Trim your hedges so they are no more than waist high. Trim tree limbs if they can be used to gain entry. Trim bushes so burglars cannot hide behind them. All doors should be visible from the street.

Garage doors and overhead doors should be checked to see if they can be opened enough to let a burglar underneath. Pad-lock garage doors with locks that have a 3/8 inch shackle. Buy the lock type that the key does not release until the lock locks. Always lock up at night and when away from home.

Remember that the weakest link is the one burglars will use. Lock up your ladder. Secure your doggie door so it cannot be used by a burglar. Post warning signs. Double check your perimeter to make sure there are no unprotected areas.

Check the condition of your fence if you have one. Clear away boxes and anything stored near your fence that can aid a burglar to climb over your fence.

All points of entry or possible break in areas should be lit. Have adequate lighting inside and outside. Protect blind alleys. Gates should be locked and in good repair. Your driveway should be visible. Secure all switch boxes on your building. Remove your name from your street address sign and mail box. Post your address in plain view and have it lit if possible. Don't hide keys outside. Give them to a trusted neighbor instead.

Inside the Home

Have adequate key control. Make sure you can separate your auto key from your house key for parking attendants and auto repair people. Remove identification from all keys in case you lose them. If you loose your keys and I.D., replace or rekey the locks at once. Know where all the keys to your home and office are. Have adequate emergency lighting inside.

Use a protective burglar alarm or intrusion device. Check it's batteries. If its over 5 years old, think about upgrading it. Make sure it covers all exterior openings in your house. It should be well maintained and tested regularly. It is best if the manufacture is still in business. The master unit should be in an out-of-the-way place like a closet.

Use a safe for protection of personal items. It should be fire and burglar resistant and securely fastened. Know its location and its contents.

Make your bedroom a nighttime security closet or retreat by adding a phone, strong lock and alarm. Have a security closet with a solid core door, pinned hinges and a dead bolt lock.

Make sure you can turn on lights without first having to walk through a dark area. It is best to have light switches at both the top and the bottom of the stairs.

Don't ever store items on stairways and keep your stair coverings in good repair.

Other Safety & Security Recommendations

Require identification from repairmen and public servants. Take down license numbers of suspicious vehicles. Buy and use timers for lights. Use locks and alarms. Close the garage door when you leave. Close drapes after dark. Use only your initials in the phone book. Don't leave messages on your door. Use a safe deposit box. Make out an operation ID list and inventory. Inventory list all property with the serial numbers recorded. Mark or etch all valuables with your driver's license number and state. Avoid display of valuables and don't show off valuables.

Place emergency numbers by the phone. Have at least one phone located where it can be reached without standing.

If you have a gun, know how to use it. Keep guns in a safe place and in good condition.

Arrive with keys in hand ready to open your door. Always lock your car door. Know your neighbor's phone number. Keep your excess cash and valuables in a bank not at home.

Tell your family what to do in case of a burglary (call police from a neighbor's phone and don't move anything). Know what to do if they discover a burglar breaking in or one already inside your home. Call police when you see something suspicious.

Small rugs and runners should have slip resistant material backing. Don't place heavy objects on high shelves. Move heavy mirrors and framed pictures located over beds, couches and chairs to a place where they won't fall on someone during a storm or earthquake. Any cupboard containing breakables should have a secure latch.

Use power tools only in well lighted areas. Keep all volatile liquids tightly capped; vapors may be toxic. Anchor top-heavy free standing furniture. Know how to shut off your water, electricity and gas. Keep the proper tools handy. Strap your water heater to a wall with metal strapping tape.

Keep necessary prescription medications and a copy of the prescription with your medical kit. Buy a portable lock and alarm for traveling. Know the location of the nearest emergency medical facility.

Appendix D

EMERGENCY ITEMS CHECKLIST

Fire Prevention Checklist

Some fire departments report as much as 15% to 30% reduction in dwelling fires or loss of life as a result of undertaking a program of home inspections.

Make a fire plan as described in the fire and natural disaster chapter and run a practice drill. Fire plans include: training, selecting a meeting place, specifying two unobstructed escape routes from every room in your house. Practice your fire plan at least once a year.

Have smoke alarms on every level of your house. If alarms are already installed, check to insure they are in good operating condition. Alarms should be placed near the bedrooms on the ceiling or 6-12 inches below the ceiling on the wall. Do not place them near air vents. Maintain them well and test them periodically.

Have fire ladders for every second and third story bedroom that has a window. Have a fire extinguisher for every hazardous area (e.g., kitchen, garage and car). Make sure you have the right type of extinguisher for the particular fire hazard area and at least two for every home. Store an extinguisher where you can get it when a fire breaks out and not where you would have to go through a fire to obtain it. For example, store the kitchen extinguisher in a room next to your kitchen.

Store a garden hose. Clean attics, garages, closet, and basements. Keep them free of rubbish. Throw out oily rags used for cleaning or store them in a protective container in a well ventilated room.

Refuel gas engines outside. Don't store gasoline, benzine or other flammable cleaning goods. If you have to store them, put them in proper metal storage containers. It is possible for the fumes of gasoline to travel as much as 10 feet and still be ignited even from a hot water heater. Don't store fireworks at all.

Check your chimney for clogs made of leaves and other debris and have your chimney cleaned often. Keep a screen in front of your fireplace. Also burning wood can cause a build up of flammable tar. For roof protection from chimney and other sparks, put on a chemical roof covering that is fire retardant.

Keep curtains away from heaters. Check your water heater for leaks and carbon build up in the burner. Check all heating sources, faulty furnaces and stoves for leaks. Replace cracked or rusty furnace parts. Be sure everything is clean and in good working order. Whenever you suspect a gas leak, call the gas company, open your windows, and get out. Keep water heater at 120 degrees or less; this is the low setting on your heater if it is not marked in degrees.

Keep matches out of the reach of children. Teach children about the dangers of fire. Don't leave children unsupervised.

Buy fire retardant clothing for pajamas. Tell smokers to always use ashtrays not trash cans. Never smoke in bed and don't leave cigarettes unattended.

In the medicine cabinet keep only the medicines currently used and the ones that have not expired. Have a phone and your shoes near your bed.

Operation Edith (Exit Drills in the Home)

Operation EDITH was developed by The National Fire Protection Association and the Fire Marshals Association and supported by The National Commission on Fire Prevention.

First, make a floor layout of your home.

1 Make an outline of the entire floor area of your home.

2 Draw in rooms and label them.

3 Locate all possible exits from each room. Include windows, doors, stairs, roofs and anything that could be used as a fire escape.

Second, do a room by room inspection.

1 Select the two best exits from a fire.

2 Test each exit. For windows and screens make sure they open easily, are low enough and offer enough crawl space for a human to use as an exit.

Third and final, complete the escape plan.

Use arrows to show the normal exit from each room on your floor layout. Draw arrows to show the emergency exit in case a fire blocks the normal exit.

Electrical Checklist

Inadequate household wiring conditions are when lights dim, fuses blow often and the television acts funny.

Make sure your bulbs do not exceed the wattage rating of the light fixture. Too high a wattage can lead to a fire by overheating. If the fixture does not indicate a maximum wattage it is best not to go beyond 60 watts.

Check your fuse box or circuit breaker box. Make sure fuses are the proper type and right size. Keep extra fuses handy. Use only the proper fuse in your circuit breaker box. Every year it is wise to test your circuits by turning them off and on ten times for each circuit. Don't overload a circuit because the hot wires inside your wall may cause a fire.

Electrical cords should not be placed in walking areas. Don't put extension cords under rugs. Make sure furniture is not resting on cords. Unwrap all cords that are being used. Don't attach cords with nails or staples. Electrical cords should only be used temporarily. Check for frayed or cracked electrical cords. Buy new ones if frayed at all. Cords should not carry more than their rated load.

If your outlets are warm to the touch have an electrician check them for faulty wiring. Make sure plugs fit snugly into their outlets. If an outlet does not have a face plate and wiring is exposed, install a face plate!

Keep all equipment cords in good condition and replace them if frayed or cracked. Have equipment placed where air can circulate around it. Only buy portable heating equipment that has a seal of a nationally recognized testing laboratory. Heaters must be placed far away from combustibles like drapes and newspapers. Make sure heaters are in a stable location so they won't be tipped over. If equipment uses a three prong plug, use only a three hole socket; the third prong is for proper grounding and is important for your safety. Don't try to defeat them by using adapters, extension cords or by cutting off the third prong.

Kitchen appliances should be unplugged when not in use. Make sure appliance cords do not come near any hot surface, (e.g., an oven or toaster) or water.

If you ever get a slight shock from anything electrical, use this as a certain warning that something is wrong and have an electrician check it out. Washers and dryers that have an excessive vibrating movement should be leveled and, if they still shake, have an appliance dealer look at them.

Keep all combustibles away from your range top, (e.g., plastic utensils, potholders, curtains, and towels). When cooking, wear clothing with short or close fitting sleeves. Where electrical shock hazard is high, for instance in kitchens and bathrooms, consider installing ground fault circuit interrupters. They can prevent many electrocutions. They are the wall plugs and circuit breakers with the reset buttons. If you have this type, check them at least once a year.

Unplug bathroom electrical appliances when not being used. Portable heaters are a higher than normal risk in bathrooms. Because of water and many grounded surfaces, I recommend you do not use portable heaters in the bathroom.

Make sure electrical blankets are in good condition. Check for breaks in wiring and plugs. Check for charred spots on either side. Don't have anything covering an electric blanket when its being used, this includes household pets. Keep electrical blankets flat and do not fold them back over themselves. Do not fall asleep when using a heating pad. Many pads can cause burns even at relatively low settings.

All outside plugs should have waterproof covers. Christmas trees are dangerous when dried out. When they become dried out remove the lights.

Automobile Checklist Items to Take Along

Get your car serviced regularly and keep it in good running condition.

Items to keep in your vehicle:

Check to see if you have all of these.

1) Small fire extinguisher

2) First aid kit

3) Alarm system

4) Blanket

5) Flares and/or emergency light

6) Flashlight

7) Pen and pad of paper

8) Change for an emergency phone call

9) Insurance record with the policy number and phone number

10) One days nourishment (health food bars)

11) Tool kit or at least enough tools to change a flat tire

12) Multipurpose campers knife

13) Emergency flares

14) A white cloth

15) A extra quart of oil

16) Special radiator hose tape

17) Store antifreeze/coolant and windshield cleaner fluid

18) For the car in winter: snow chains (tarpaulins help to put them on), container of sand, shovel, windshield scrapper, tow chain or rope, heavy gloves, over shoes, extra woolen socks, and winter head gear are good to keep in the trunk.

19) Install an auto alarm, kill switch, ignition and accessory locks.

When traveling during the winter, stay up on the current weather conditions by listening to public radio stations or a special weather channel usually 162.55 or 162.4 MHz. If you get stuck in a storm, don't panic. Don't try to walk out, stay with your car. It easy to become disorientated. Don't try to dig or push yourself out if you are by yourself because over exertion lowers your resistance to the cold. Tie a cloth to your radio antenna. Keep the snow off the exhaust and air vents near the windshield; this measure ensures exhaust fumes won't build up inside the car. Run the engine sparingly and don't let the windows freeze shut. Keep warm and rub your arms and legs for increased circulation. If it is night time, leave the inside light on and make sure someone stays awake to watch for help.

In rural areas or if you do a lot of traveling, have an emergency citizen band radio. Use operation I.D. on car accessories and parts and don't forget to fill out the included door insert card.

Natural Disaster Checklist

Know where your main shut offs are for water, gas and electricity. Store a pipe and crescent wrench to turn them off. Know where your nearest civil defense shelter is located.

These supplies will help you if you stay home during a disaster by reducing the hardships. Take the supplies with you if you go to a shelter to use en route. They also might help you when you get there. Things to store at home in case of emergency are:

1) A good first aid kit with large bandages.

2) A week's worth of dried foods for each person. Use the type that does not require refrigeration.

3) Canned goods and an opener.

4) Powdered milk.

5) A filled water container; store 3 gallons for each person.

6) Canned or bottled drinks with fruit and vegetable juices. You can use the fluids from canned foods.

7) Water purifying tablets.

8) You can purify water in an emergency by adding 8 drops of liquid bleach (bleach being a 5% hypochlorite solution) to a gallon of water; use 16 drops if the water is cloudy, stir it and mix it. Sometimes during a emergency you can store water in a clean bathtub or use the 50 gallons stored in your hot water heater.

9) Blanket or sleeping bag for each member of family.

10) Medicines required by family members.

11) Pad of paper with pen to take notes.

12) Plastic bags for human waste and powdered chlorine bleach to disinfect sewage.

13) Lantern and flashlight with fresh batteries.

14) A battery powered radio.

15) A covered container to use as a toilet.

16) Emergency cooking equipment.

17) Cooking and eating utensils, clothing, bedding, general equipment and tools.

18) Fifty feet of strong utility rope. Miscellaneous items like matches, candles, and personal convenience items.

19) In winter store an adequate supply of heating fuel.

20) At work keep a pair of comfortable shoes, a day's nourishment (health bars) and a small flashlight.

In areas of the country subject to hurricanes or floods, it is also wise to keep on hand emergency materials needed to protect your home like plywood sheeting or lumber to board up windows and doors and plastic sheeting to protect furniture. Sand bags are good unless flooding is imminent then they may cause structural damage if used.

Baby-sitting Safety Tips

It is important to be selective when choosing a person with whom you plan to leave your children. Many children have been physically injured and/or sexually abused by people who were hired to "take care" of them.

Tips for choosing a sitter:

1. Record all information about the baby-sitter. This includes: full name, address, drivers license number, type of car and license plate number, and information on the baby-sitter's family

2. Make sure the baby-sitter is adequately trained and has a good sense of responsibility and maturity. A baby-sitter trained in first aid is desireable. A local Red Cross teaches these things.

3. If you are going leave you child at the sitter's house, know the other people who will be there. Make suitable arrangements for seeing the sitter's house.

Tips for parents of the teen-age baby-sitter:

1. Know something about the people for whom your son or daughter is baby-sitting.

2. Have your son or daughter phone periodically so you'll know everything is all right.

3. Always check with your son or daughter about the job and how it went.

4. Make sure transportation has been arranged.

5. Either wait up for your child or have them wake you up as soon as they arrive home.

6. Set reasonable hours beyond which your son/daughter may not baby-sit.

Tips for the baby-sitter:

1. Know your employer. Do not put your name and number on bulletin boards or in the paper indiscriminately when seeking baby-sitting jobs. Look into each situation carefully.

2. Make arrangements for transportation beforehand, both to and from the job. Do not walk home alone especially at night.

Appendix E

OPERATION IDENTIFICATION

Introduction

Operation Identification can dramatically cut burglary rates! In Phoenix Arizona, people who didn't join Operation ID had 18 times as many burglaries as their neighbors who did join. It is reported this plan has reduced burglary up to 75%. Instituted in Monterey Park, California in 1963, after a few years only 3 out of 4,000 participants reported burglaries compared to a nonparticipating sample that had 1,800 burglaries out of 7,000 residences.

How You Can Participate

Operation ID consists of you marking all your valuables and making an inventory sheet including all household items, car items, personal items and credit cards in order to facilitate recovery in the event they are stolen. Mark your valuables by etching your driver's license number on them and marking cloth items, like furs and rugs, with a indelible ink marking pen. Mark all valuables that might be stolen: possessions like TV's, radios, stereos, typewriters, golf clubs, appliances, tools- any portable item even if they have a serial number.

How it Works

Mark items with your state abbreviation and your driver's licence number; don't use your social security number. If you don't have a driver's license, get a state ID card from your State Department of Motor Vehicles. If you cannot do this, use your name and phone number. For large items it is best to mark the right upper hand corner and for furniture mark the bottom right hand side. Small items, fragile items and expensive items like jewelry or collections should be photographed with your drivers license number and the state of issuance (use quality film and a indoor flash, place items around the photo card provided with this kit). Furs should be marked on the inside directly on the pelt with indelible ink or invisible ink that shows only under black light. When you sell a marked item, delete the old Operation ID number by marking through the old number from the upper left to the bottom right, leaving the old number so it can be read.

Why Does it Work so Well

It is good for two reasons. It is less likely you will be robbed and more likely you will get items back if taken; it is also negative incentive for a burglar to rob your house. The burglar is aware he is more likely to get caught when he tries to sell marked goods. Possession of stolen goods is a crime and marked goods are easy to trace. Thieves are detered from taking marked property because it is harder to hawk or pawn and it brings a lower price. Marking your property shows you are aware of and are concerned about crime. If your property is stolen, you are more likely to get items back because markings identify you as the owner. Marked property is less attractive to thieves and once stolen it is more likely to be returned to you by the police.

Using a door decal shows to burglars you are aware of the crime potential and you take precautions to prevent crime. A decal tells potential burglars to stay away. You aren't likely to see your possessions again after a robbery; in fact, you will not get anything back if nothing identifies you as the owner. Cities auction off millions in dollars worth of unmarked property every year because they cannot locate the owner.

HOUSEHOLD INVENTORY AND INSURANCE EVALUATION LIST

(PAGE 2)

HOUSE HOLD INVENTORY AND INSURANCE EVALUATION (PAGE 1)

Name
Address
Date of Inventory

Revised
Revised
Revised

By creating a permanent record of your personal property you will be in a better position in the event of a burglary or natural disaster loss. It will help recover your property or report your insurance losses adequately.

Complete this list at once and keep the list in a safe place out of your home. In a fire or burglary you could lose it. You can place a copy in a safe deposit box or mail one to a relative. If you suffer a loss, contact your agent at once and take precautions to protect property from further damage. This list is valuable in determining your insurance needs; to insure in this era of inflation is to your benefit.

If you just cannot find the time to complete this list, you should make a video tape of your house or hire someone to do it for you. You can carry around an audio cassette machine and create a list verbally, or at least take still photographs of each room in its natural state. Mark the back of the photos with the date and the room. It is a good idea to take photos anyway to refresh your memory in addition to the written inventory. You can take your chances by not doing anything and probably get far less after a fire or break in.

Photograph all expensive valuables next to a photo-card included in this book to supplement the list.

RECORD OF INSURANCE POLICY

Policy No.	Insurance Company	Property Covered	Premium Amount	Expiration Date

Person who has wills (caution: do not put wills in a safe deposit box)

Name:

(PAGE 3)

Cars and vehicles (motorcycles, rv's, trailers, boats, etc.)

Type	Age	Color	Lic No.	Vin No.*

*Vin - vehicle identification number, copy off the metal plate on drivers side of the dashboard near the windshield.

Social security numbers of entire family

CREDIT CARD RECORD

Name of Issuing Company	Address	Phone	Card No.

Savings and checking accounts name and branch of bank

Saftety deposit box number, bank, and branch

Stocks and Bonds

No. of	Name	Shares

LIVING ROOM (PAGE 4)

Description	# of Pieces	Date Bought	Original Cost	Present Value	Ser No Marks
Book Cases					
Brick-Brac (24)					
Books (19)					
Cabinets					
Contents					
Chairs					
Clocks					
Cushions					
Davenports					
Desk					
Drapes					
Fire Place Accs.					
Lamps					
Mirrors					
Musical Instruments					
Paintings (24)					
Piano					
Air Conditioner					
Art Photographs(24)					
Radio					
Rugs\ Carpets					
Television					
Tables					
Sofa					
Stereo					
Video Records					
Miscellaneous					
Plants					

Total

DINING ROOM (PAGE 5)

Description	# of Pieces	Date Bought	Original Cost	Present Value	SerNo Marks
Buffet					
China(25)					
China Cabinet					
Clocks					
Chairs					
Drapes					
Electric appliances					
Glassware(25)					
Lamps					
Lighting Fixtures					
Linens(20)					
Mirrors					
Painting (24)					
Rugs					
Silverware (25)					
Table					
Miscellaneous					
Air Conditioner					

Total

KITCHEN (PAGE 6)

Description	# of Pieces	Date Bought	Original Cost	Present Value	Ser No Marks
Chairs					
Clock					
Cutlery					
China & Glass(25)					
Dishwasher (port.)					
Electrical Appl.					
Food Processor					
Freezer					
Griddle					
Coffee Maker					
Roaster					
Mixer					
Refrigerator					
Toaster					
Waffle Iron					
Vacuum					
Flatware					
Lighting fixtures					
Microwave Oven					
Radio					
Rug					
Stove					
Sweepers, Mops					
Table					
Miscellaneous					
Curtains					
Exhaust Fan					
Foodstuffs					
Floor Waxer					
Pots & Pans					
Wines & liquors					

Total

HALL (PAGE 7)

Description	# of Pieces	Date Bought	Original Cost	Present Value	Ser No Marks
Cabinets					
Clocks					
Curtains					
Chairs					
Drapes					
Lamps					
Mirrors					
Pictures (24)					
Rugs					
Tables					
Miscellaneous					
Total					

DEN

Description	# of Pieces	Date Bought	Original Cost	Present Value	Ser No Marks
Chairs					
Clocks					
Davenport					
Drapes					
Lamps					
Mirrors					
Pictures(24)					
Radio					
Rugs					
Television					
Tables					
Miscellaneous					
Totals					

BEDROOM NO 1 (PAGE 8)

Description	# of Pieces	Date Bought	Original Cost	Present Value	Ser No Marks
Bedding(20)					
Bed					
Box Spring					
Bureaus					
Chairs					
Chest					
Clocks					
Desk					
Drapes					
Dresser					
Dressing Table					
Lamps					
Mattresses					
Mirrors					
Night Stand					
Pictures (24)					
Radio					
Television					
Toiletries					
Miscellaneous					
Rugs					
Air Conditioner					

Total

BEDROOM NO 2 (PAGE 9)

Description	# of Pieces	Date Bought	Original Cost	Present Value	Ser No Marks
Bedding(20)					
Bed					
Box Spring					
Bureaus					
Chairs					
Chest					
Clocks					
Desk					
Drapes					
Dresser					
Dressing Table					
Lamps					
Mattresses					
Mirrors					
Night Stand					
Pictures (24)					
Radio					
Television					
Toiletries					
Miscellaneous					
Rugs					
Air Conditioner					

Total

BEDROOM NO 3 (PAGE 10)

Description	# of Pieces	Date Bought	Original Cost	Present Value	Ser No Marks
Bedding(20)					
Bed					
Box Spring					
Bureaus					
Chairs					
Chest					
Clocks					
Desk					
Drapes					
Dresser					
Dressing Table					
Lamps					
Mattresses					
Mirrors					
Night Stand					
Pictures (24)					
Radio					
Television					
Toiletries					
Miscellaneous					
Rugs					
Air Conditioner					

Total

BEDROOM NO 4 (PAGE 11)

Description	# of Pieces	Date Bought	Original Cost	Present Value	Ser No Marks
Bedding(20)					
Bed					
Box Spring					
Bureaus					
Chairs					
Chest					
Clocks					
Desk					
Drapes					
Dresser					
Dressing Table					
Lamps					
Mattresses					
Mirrors					
Night Stand					
Pictures (24)					
Radio					
Television					
Toiletries					
Miscellaneous					
Rugs					
Air Conditioner					
Total					

BATHROOM NO 1 (PAGE 12)

Description	# of Pieces	Date Bought	Original Cost	Present Value	Ser No Marks
Chairs					
Chest					
Drapes					
Electrical Apply.					
Hair Dryer					
Heat Lamp					
Razor					
Lamps					
Linens (20)					
Medicines					
Mirrors					
Rugs					
Table					
Scale					
Vanity					
Hamper					
Total					

BATHROOM NO 2

Description	# of Pcs.	Date Bought	Original Cost	Present Value	Ser No Marks
Chairs					
Chest					
Drapes					
Electrical Apply.					
Hair Dryer					
Heat Lamp					
Razor					
Lamps					
Linens (20)					
Medicines					
Mirrors					
Rugs					
Table					
Scale					
Vanity					
Hamper					
Total					

BATHROOM NO 3

(PAGE 13)

Description	# of Pieces	Date Bought	Original Cost	Present Value	Ser No Marks
Chairs					
Chest					
Drapes					
Electrical Apply.					
Hair Dryer					
Heat Lamp					
Razor					
Lamps					
Linens (20)					
Medicines					
Mirrors					
Rugs					
Table					
Scale					
Vanity					
Hamper					
Total					

BATHROOM NO 4

Description	# of Pcs.	Date Bought	Original Cost	Present Value	Ser No Marks
Chairs					
Chest					
Drapes					
Electrical Apply.					
Hair Dryer					
Heat Lamp					
Razor					
Lamps					
Linens (20)					
Medicines					
Mirrors					
Rugs					
Table					
Scale					
Vanity					
Hamper					
Total					

ATTIC (PAGE 14)

Description	# of Pcs.	Date Bought	Original Cost	Present Value	Ser No Marks
Chairs					
Luggage (17)					
Pictures (24)Tables					
Stored Items					
Total					

LAUNDRY ROOM

Description	# of Pieces	Date Bought	Original Cost	Present Value	Ser No Marks
Baskets					
Chairs					
Clock					
Dryer					
Iron Board					
Laundry Supplies					
Linens (20)					
Radio					
Stove					
Tables					
Tub					
Washer					
Total					

RECREATION ROOM (PAGE 15)

Description	# of Pcs.	Date Bought	Original Cost	Present Value	Ser No Marks
Bar & Equip.					
Chairs					
Cabinets					
Clocks					
Davenports					
Drapes					
Fireplace Equip.					
Games					
Liquors					
Pictures (24)					
Radio					
Tables					
Television					
Trophies					
Miscellaneous					

Total

BASEMENT (PAGE 16)

Description	# of Pieces	Date Bought	Original Cost	Present Value	Ser No Marks
Food Supplies					
Fuel					
Hand Tools					
Household Supplies					
Mops					
Sweepers					
Hardware					
Power Tools					
Work Bench					
Total					

JEWELRY AND FURS (PHOTOGRAPH WITH PHOTO-CARD)

Description	# of Pieces	Date Bought	Original Cost	Present Value	Ser No Marks
Watches					
Rings					
Total					

GARAGE (PAGE 17)

Description	# of Pieces	Date Bought	Original Cost	Present Value	Ser No Marks
Automobiles					
Auto Accessories					
Auto Tools					
Christmas Items					
Garden Tools					
Hose					
Lawn mower					
Patio Furniture					
Snow Remover					
Work Bench					
Total					

LUGGAGE

Description	# of Pieces	Date Bought	Original Cost	Present Value	Ser No Marks
Total					

SPORTS AND HOBBY EQUIPMENT (PAGE 18)

Description	# of Pieces	Date Bought	Original Cost	Present Value	Ser No Marks
Guns					
Total					

BOOKS (PAGE 19)

Description	# of Pieces	Date Bought	Original Cost	Present Value	Ser No Marks

Total

LINENS AND BEDDING (PAGE 20)

Description	# of Pieces	Date Bought	Original Cost	Present Value	Ser No Marks

Total

MEN'S CLOTHING (PAGE 21)

Description	# of Pieces	Date Bought	Original Cost	Present Value	Ser No Marks

Total

WOMEN'S CLOTHING (PAGE 22)

Description	# of Pieces	Date Bought	Original Cost	Present Value	Ser No Marks

Total

CHILDREN'S CLOTHING (PAGE 23)

Description	# of Pieces	Date Bought	Original Cost	Present Value	Ser No Marks

Total

PICTURES & ART WORK (PAGE 24)

Description	# of Pieces	Date Bought	Original Cost	Present Value	Ser No Marks

Total

CHINA, GLASS, & SILVERWARE (PAGE 25)

Description	# of Pieces	Date Bought	Original Cost	Present Value	Ser No Marks

Total

MISCELLANEOUS (PAGE 26)

Description	# of Pieces	Date Bought	Original Cost	Present Value	Ser No Marks

Total

MISCELLANEOUS (PAGE 27)

Description	# of Pieces	Date Bought	Original Cost	Present Value	Ser No Marks

Total

MISCELLANEOUS (PAGE 28)

Description	# of Pieces	Date Bought	Original Cost	Present Value	Ser No Marks

Total

(PAGE 29)

SUMMARY	PAGE	ORIGINAL	PRES. VALUE
ATTIC	14		
BASEMENT	16		
BATHROOMS 1 & 2	12		
BATHROOMS 3 & 4	13		
BEDROOM #1	8		
BEDROOM #2	9		
BEDROOM #3	10		
BEDROOM #4	11		
BOOKS & RECORDS	19		
CHINA, GLASS, & SILVER	25		
CLOTHING			
CHILDREN'S	23		
MEN'S	21		
WOMEN'S	22		
DEN	7		
DINING ROOM	5		
GARAGE	17		
HALLS	7		
JEWELRY & FURS	16		
KITCHEN	6		
LAUNDRY ROOM	14		
LINENS	20		
LIVING ROOM	4		
LUGGAGE	17		
MISCELLANEOUS	26-28		
PICTURES & BRICK-A-BRACK	24		
RECREATION ROOM	15		
SPORTS & HOBBY EQUIPMENT	18		

TOTAL PRESENT VALUE

PRESENT VALUE - PRESENT INSURANCE COVERAGE = ADDITIONAL COVERAGE NEEDED

Emergency Work Sheet (Cut Out)

Fill this sheet out and discuss it with all family members and relatives. Post this in a central place in your home (e.g., near the phone in the living room).

In the event of a large emergency, all family members will meet at shelter listed below:

LOCATION:

Any family member leaving the meeting place before others will leave a message at:

LOCATION:

If there is a serious medical problem, we will first try to go to:

NAME OF HOSPITAL:

ADDRESS:

If we are un-able to go to that hospital, we will try:

NAME OF HOSPITAL:

ADDRESS:

Location of nearest fire station:

LOCATION:

We will leave phone messages, if not reunited within 24 hours, at this out-of-the area location: Ask someone out-of-the area if you can use them as a phone relay center (avoid using the phone in the hours right after major emergencies unless it is a matter of life and death).

NAME:

PHONE:

Know the evacuation and emergency procedures of your school and work place. Many times schools will keep children till a parent picks them up. Also, some key employees will be expected to stay on the job.

WALLET CARD (CUT OUTS)

WALLET CARD
ADDRESS
PHONE
NEGH. PH.
LIC. PLATE
VIN. NO.*
INSUR. PH#
POLICY NO.
DOCTOR
RELATIVE
CAR TYPE
BODY STYLE
COLOR
ENGINE SIZE

WALLET CARD
ADDRESS
PHONE
NEGH. PH.
LIC. PLATE
VIN. NO.*
INSURANCE PH#
POLICY NO.
DOCTOR
RELATIVE
CAR TYPE
BODY STYLE
COLOR
ENGINE SIZE

BIKE WALLET CARD
SERIAL NO.
MAKE
MODEL
GEARS
COLOR
VALUE
LOCK KEY NO.
DESCRIPTION

*VIN VEHICLE IDENTIFICATION NUMBER/ ON THE DASHBOARD ON THE DRIVERS SIDE NEXT TO WINDSHIELD

SKI WALLET INSERT (CUT OUTS)

CARRY WITH YOU WHEN SKIING AND STORE IN SKI BOOTS AFTER YOUR TRIP

SKI WALLET INSERT

MAKE MODEL

COLOR SERIAL NO.

IDENTIFYING MARKS

BOOTS

MAKE MODEL

COLOR SERIAL NO.

IDENTIFYING MARKS

BINDINGS

MAKE MODEL

COLOR SERIAL NO.

IDENTIFYING MARKS

SKI WALLET INSERT

MAKE MODEL

COLOR SERIAL NO.

IDENTIFYING MARKS

BOOTS

MAKE MODEL

COLOR SERIAL NO.

IDENTIFYING MARKS

BINDINGS

MAKE MODEL

COLOR SERIAL NO.

IDENTIFYING MARKS

Vacation Mailers (Cut Outs)

SERVICE STOP
NAME
ADDRESS
CITY PHONE
THE ABOVE PERSON WANTS TO
TEMPORARILY STOP SERVICE BETWEEN
AND PLEASE SUSPEND DELIVERY

NEWSPAPER
POST OFFICE PLEASE HOLD MAIL UNTIL
MILKMAN
OTHER

POLICE

OUR RESIDENCE WILL BE VACANT FROM
UNTIL
PEOPLE WITH KEYS ARE
NAME
ADDRESS
CITY PHONE
FOR EMERGENCY CONTACT

EXPECT LIGHTS TO GO ON AND OFF YES, NO

SERVICE STOP
NAME
ADDRESS
CITY PHONE
THE ABOVE PERSON WANTS TO
TEMPORARILY STOP SERVICE BETWEEN
AND PLEASE SUSPEND DELIVERY

NEWSPAPER
POST OFFICE PLEASE HOLD MAIL UNTIL
MILKMAN
OTHER

PHOTO CARD, CAR BODY INSERTS (CUT OUTS)

PHOTO-CARD
PAPER FOR PHOTOGRAPHING VALUABLES NEXT TO

FIRST INITIAL
LAST NAME
DRIVERS LICENSE #

STATE

/ 1" / 2" / 3" / 4" /

PLACE CAR BODY INSERTS INSIDE THE CAR DOOR PANEL THROUGH THE WINDOW SLOT

CAR BODY INSERTS
NAME OF CAR OWNER
DRIVERS LICENSE NO.
MAKE (Dodge)
MODEL (Ram Charger)
CAR VIN NO. LIC. PLATE NO.
DATE

CAR BODY INSERTS
NAME OF CAR OWNER
DRIVERS LICENSE NO.
MAKE (Dodge)
MODEL (Ram Charger)
CAR VIN NO. LIC. PLATE NO.
DATE

CAR BODY INSERTS
NAME OF CAR OWNER
DRIVERS LICENSE NO.
MAKE (Dodge)
MODEL (Ram Charger)
CAR VIN NO. LIC. PLATE NO.
DATE

Instructions for the baby-sitter to follow:

(Cut Out) Date

Know emergency numbers located by the phone (fire, police, doctor, neighbors phone) check address listed near phone

Will be at:

Phone:

Will Return at:

Bedtime:

Medications:

Special Instructions:

Calls Received:

1) Do not open the door to anyone you don't know.

2) Keep all outside doors and windows locked.

3) If you have any trouble or see anything suspicious phone the police.

4) No visitors are allowed and do not leave the house unattended.

5) Keep the radio and TV on low to hear unusual noises or the baby crying. Check up on sleeping children.

6) Use the phone only if there is a emergency. Call parents if you have any questions.

7) Do not let the child leave with anyone except a parent.

8) During a fire, get everyone out fast and don't go back in.

SURVEY (Cut Out)

SHARE YOUR IDEAS WITH US SO WE CAN SHARE THEM WITH OTHERS.

NAME

ADDRESS

REASON FOR BUYING THE KIT?
WERE YOU EVER A VICTIM?

1 WHAT VALUE WAS THE BOOK TO YOU? (CIRCLE)
(FANTASTIC, GREAT, GOOD, OK, POOR)

2 DID YOU (CIRCLE ANSWERS)
FINISH THE BOOK (YES-NO)
FINISH THE INVENTORY LIST (YES-NO)
LOOK AT VACATION CHECKLIST (YES-NO)
FINISH THE WALLET INSERT CARDS (YES-NO)
FINISH THE SECURITY EVALUATION (YES-NO)
FINISH THE CAR ID PAPER (YES-NO)

3 DID YOU COMPLETE THE WHOLE PROGRAM? (YES-NO)
4 HOW DID IT RATE OVERALL? (EXEC, GOOD, OK, POOR)
5 WAS IT INFORMATIVE? (EXEC, GOOD, OK, POOR)
6 DID THE BOOK HOLD YOUR INTEREST? (YES-NO)

7 HOW DID THE KIT RATE?
CIRCLE (EXEC, GOOD, OK, POOR) RATE EACH ITEM

BOOK	(EXEC, GOOD, OK, POOR)
INVENTORY SHEET	(EXEC, GOOD, OK, POOR)
VACATION CHECKLIST	(EXEC, GOOD, OK, POOR)
SECURITY EVALUATION	(EXEC, GOOD, OK, POOR)

8 WHAT PART DID YOU LIKE MOST?
9 WHAT PART DID YOU LIKE LEAST?
10 WAS THE TEXT EASY TO READ?
11 DID THE ILLUSTRATIONS HELP?

THANK YOU, YOUR INPUT IS OF GREAT HELP TO US IN IMPROVING OUR NEXT PRINTING.

COMMENTS, SUGGESTIONS, ERRORS, WHAT WOULD YOU CHANGE OR IMPROVE? WHAT WOULD YOU ADD? PLEASE LIST ON BACK OF THIS PAGE.

Were We A Help (Cut Out)

Write us if you avoided a crime or after a crime takes place; was this kit helpful? It is important for us to know for we care about you and other people. We might be able to improve our kit or use the information you give us to help others that face similar situations. Your comments also might help police to develope their programs.

Mail To

Securityland
P.O. Box 2079
Los Gatos, CA 95031-2079

(Please print)
Name
Address
State zip
Phone (optional) ( ) -

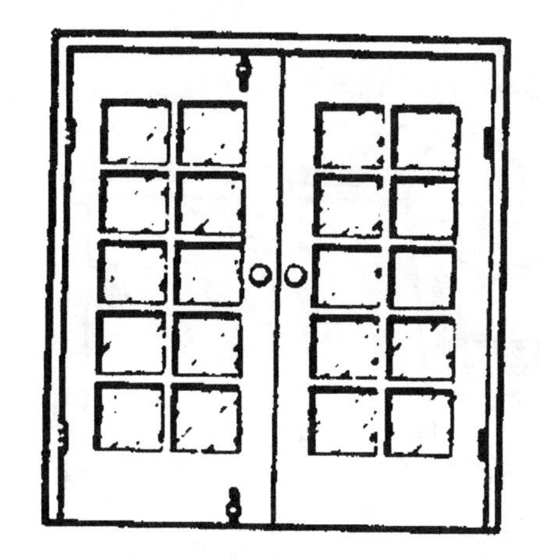

ench doors usually have glass panes but may be solid.

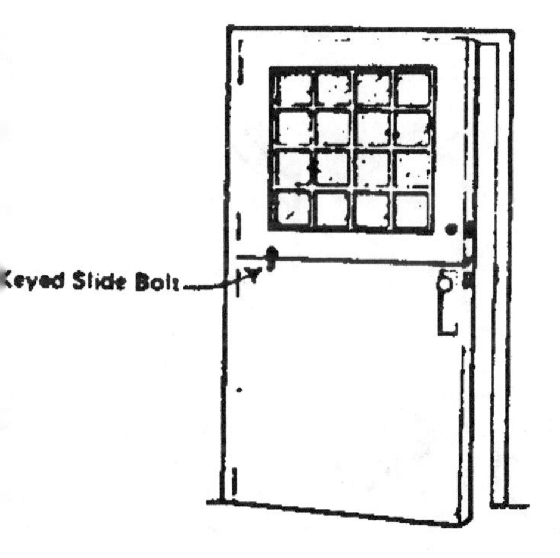

tch doors have top and bottom sections.

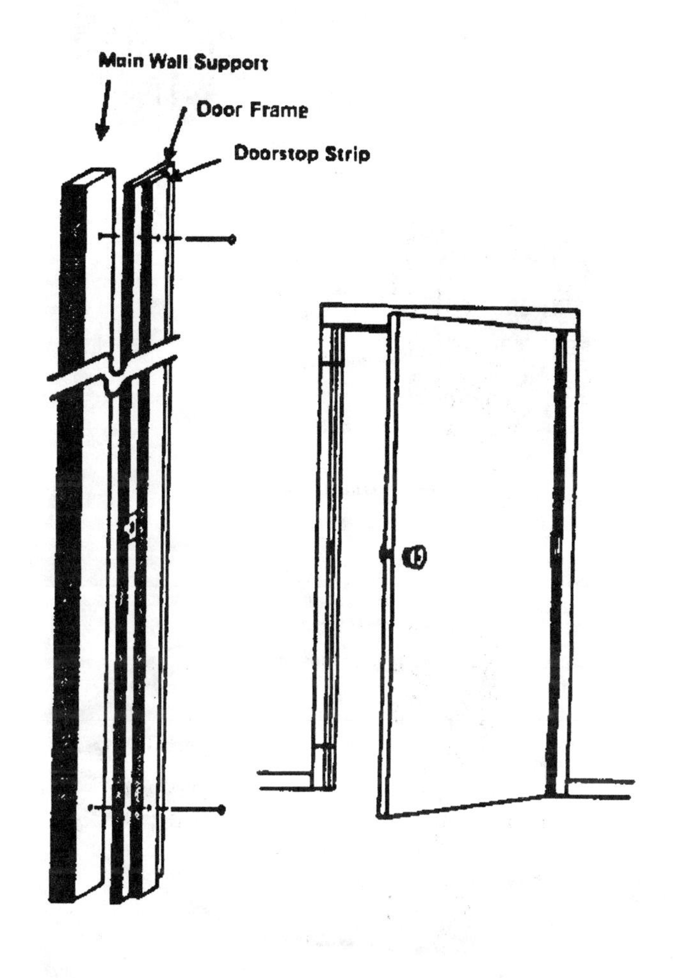

Long Screws should join the frame to the wall.

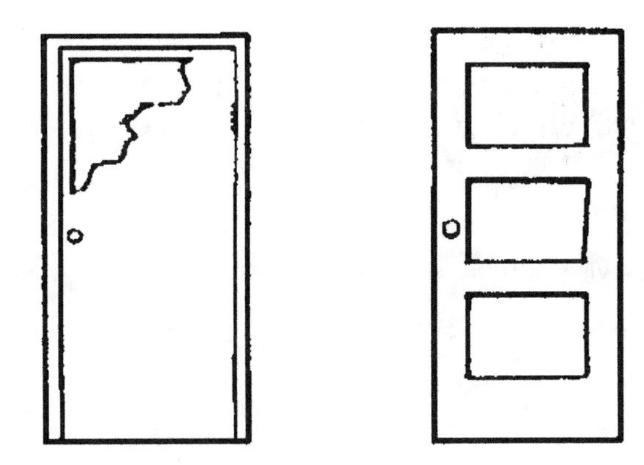

Hollow Core door is filled with corrugated cardboard and is easily broken through.

Door with thin wood panels is vulnerable and calls for a double-cylinder deadbolt.

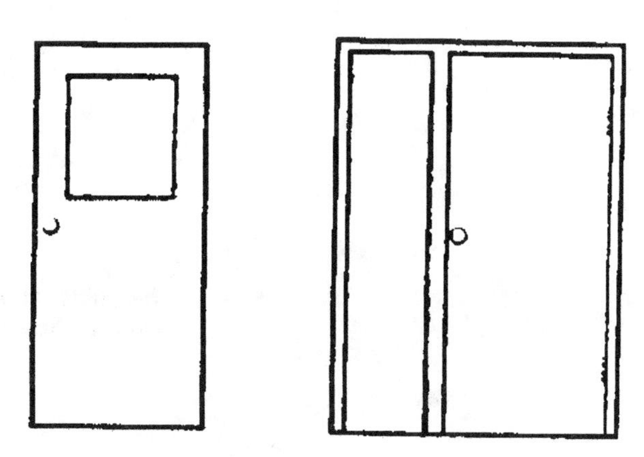

Door with glass calls for double-cylinder deadbolt.

Solid door with glass beside it calls for a double-cylinde deadbolt.

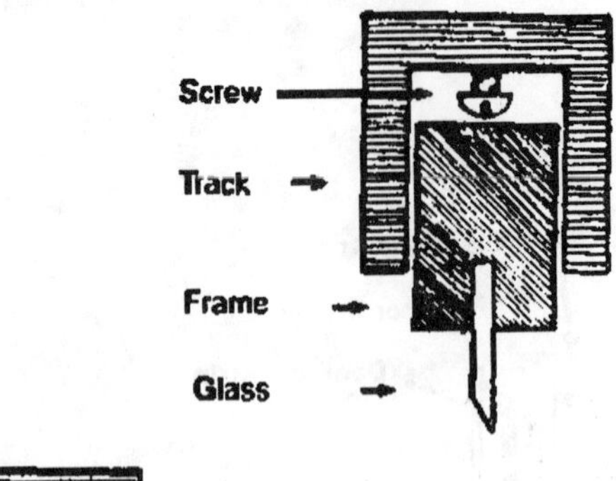

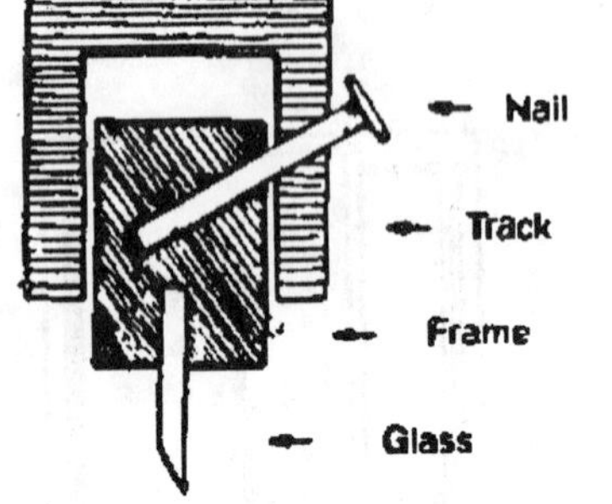

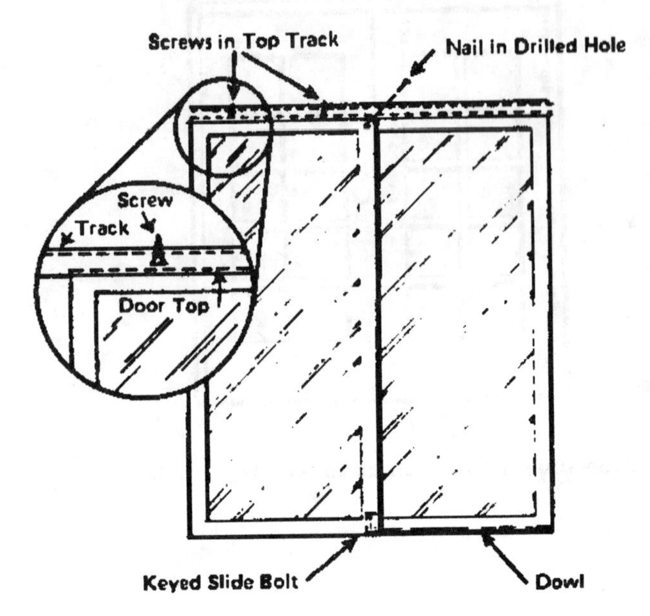

Sliding glass doors and windows are built alike.

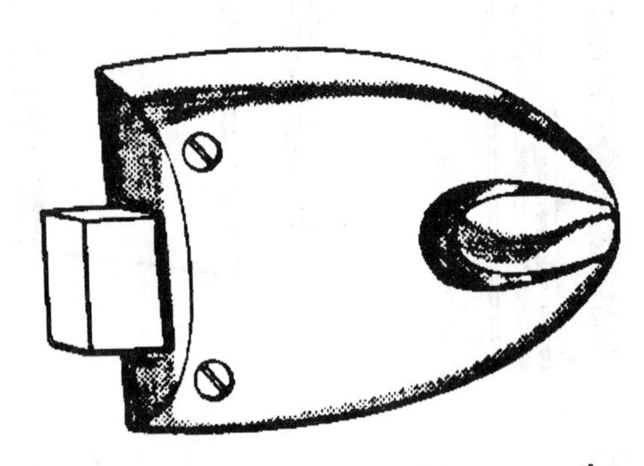

A night latch of good quality will help increase the security of a door.

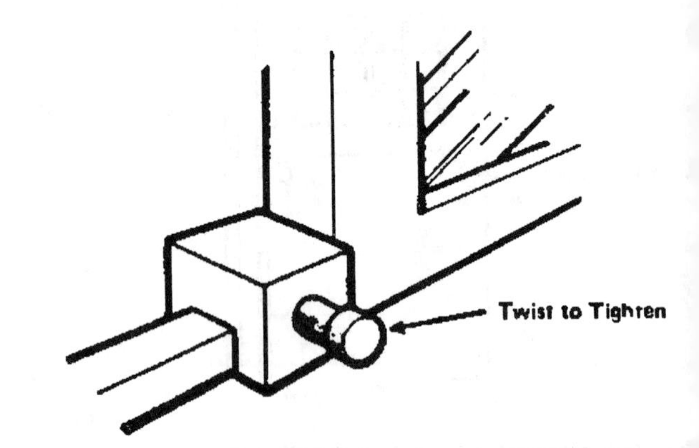

Sliding bolts for inside tracks of glass doors and windows.

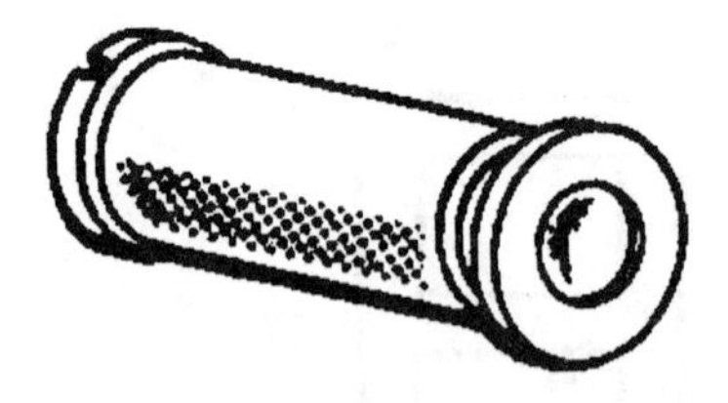

Peephole viewer permits observing who is at your door without opening the door.

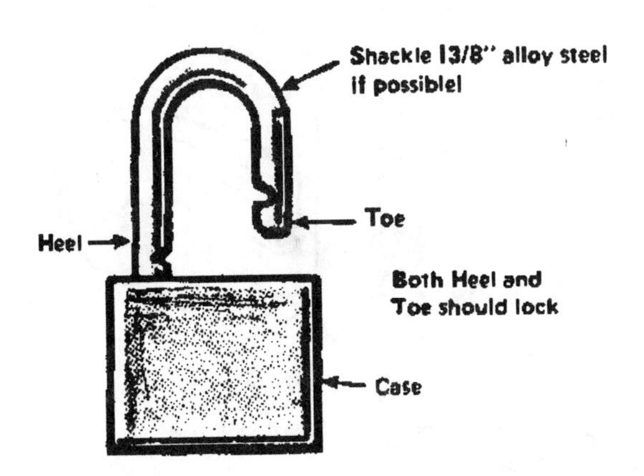

Padlock shackle should be hardened, case should be heavy.

Strike Plate is mounted on door frame.

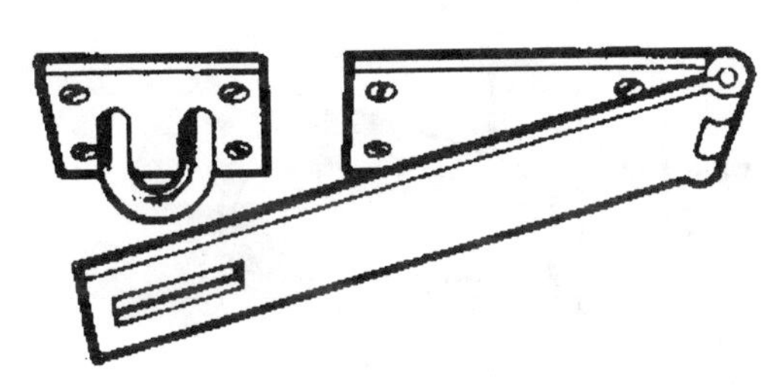

Hasp, when closed, covers screws and can't be dismounted.

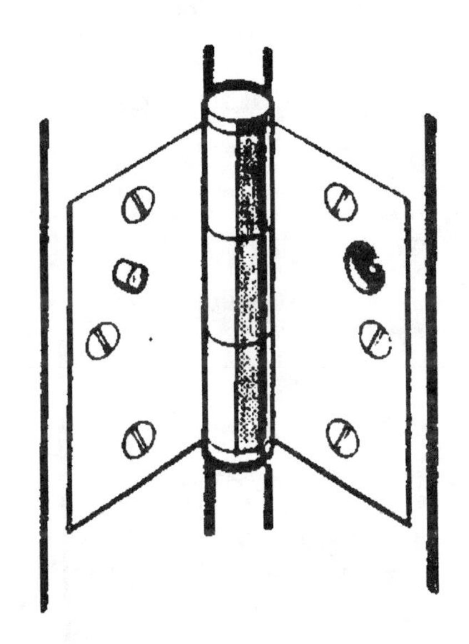

Protruding Screw and matching hole secure external hin

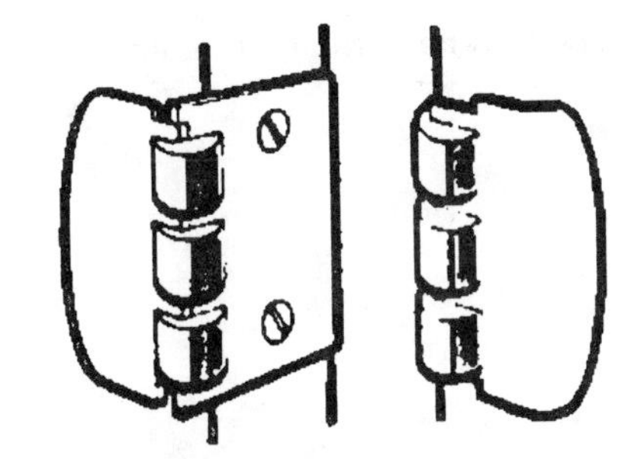

Flip Lock secures without key.

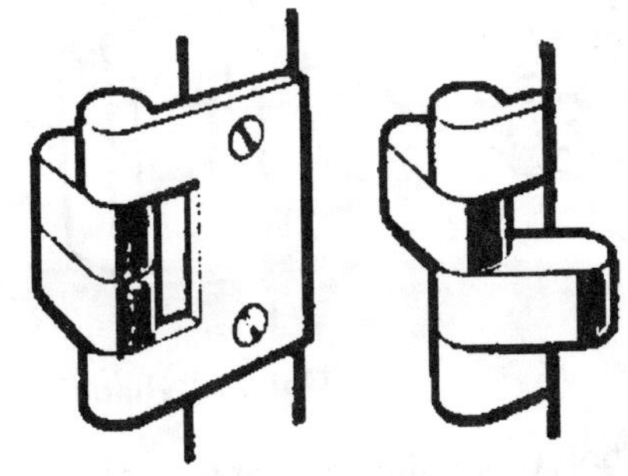

Snib Lock also locks and unlocks quickly without key.

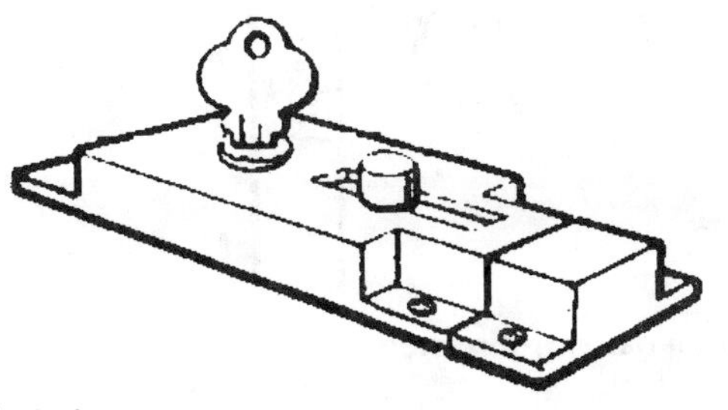

Keyed slide bolt, when locked, can't be opened without key.

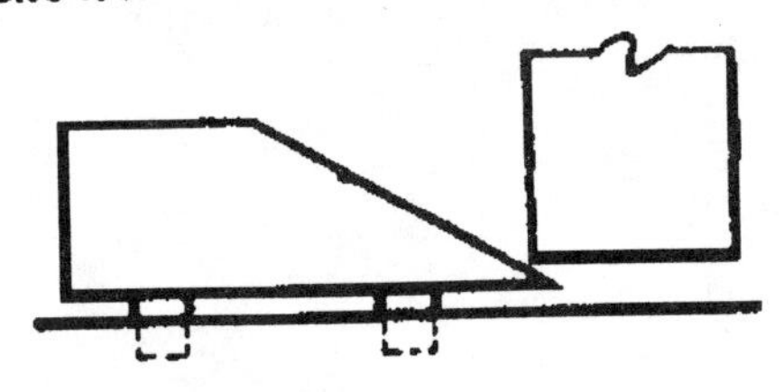

Stop Lock fits in holes in floor.

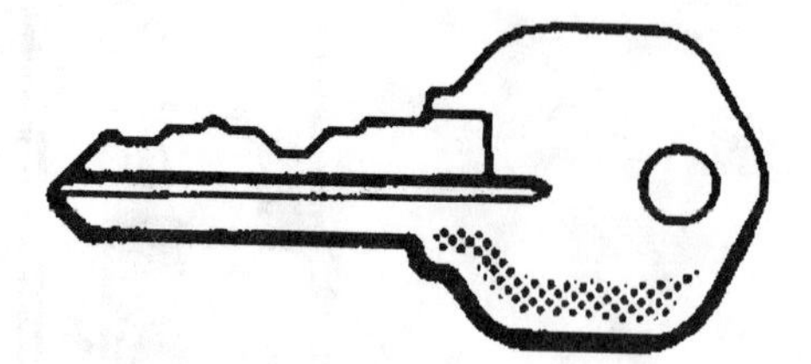

Pin Tumbler Key is common for modern locks.

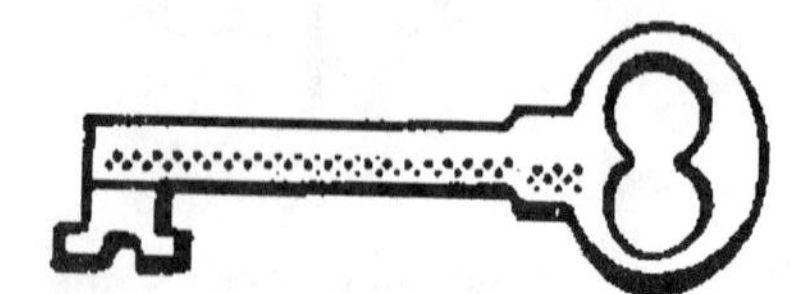

Bit Key is used on older locks.

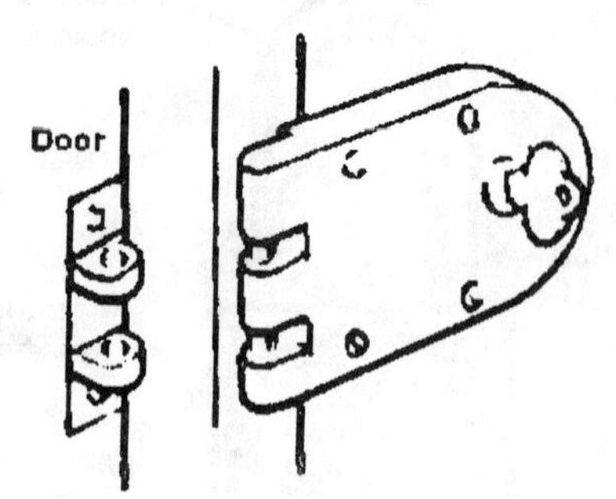

Vertical Jimmy-Resistant Deadbolt has a double cylinder model with key inside.

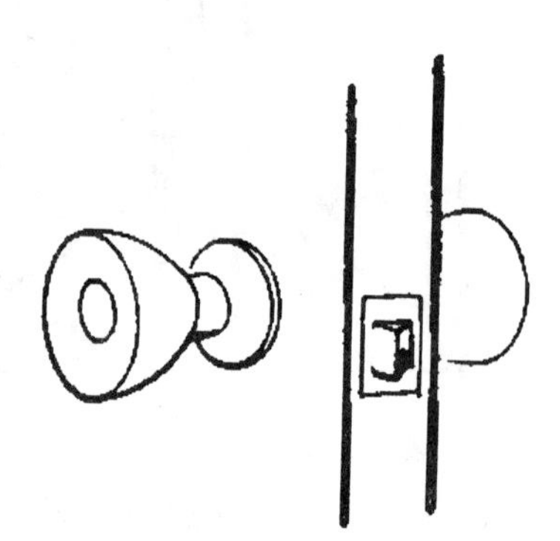

Spring Latch is easily slipped

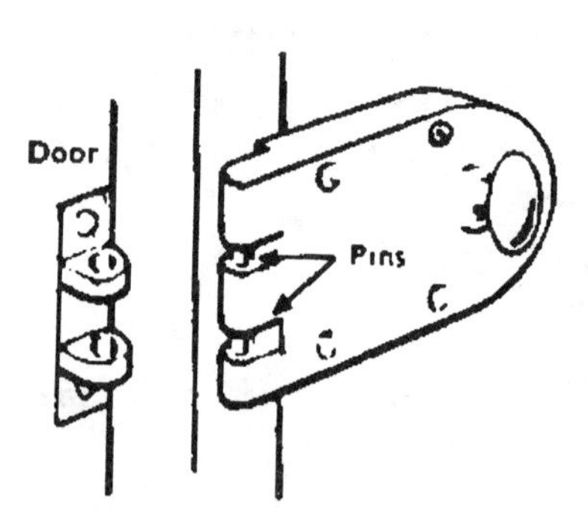

Vertical Jimmy-Resistant Deadbolt has a thumb turn inside on single cylinder model.

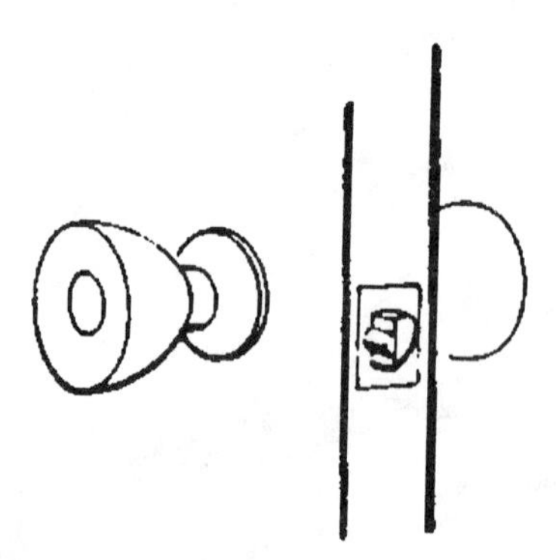

Dead Latch has anti-slip latch.

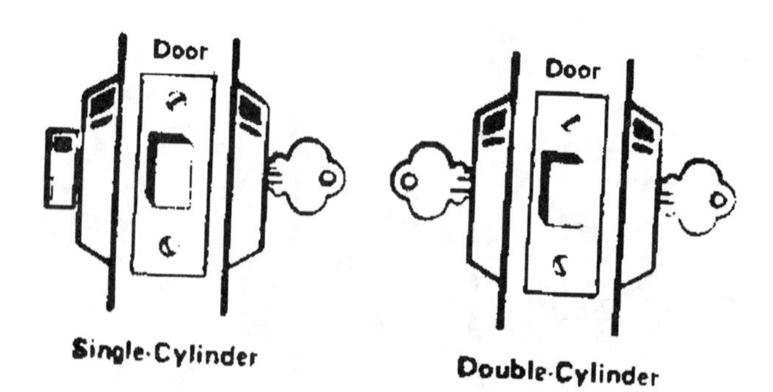

Single-Cylinder Deadbolt throw should be at least 1".
Double-Cylinder Deadbolt throw should be at least 1".

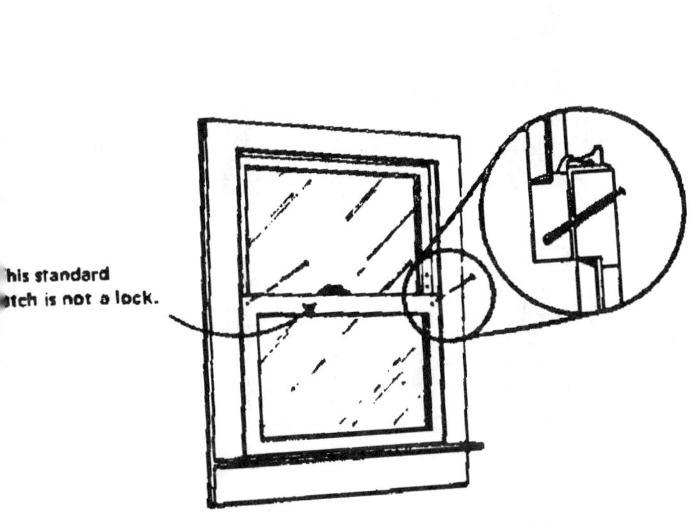

Double-hung windows have two sections with a latch between.

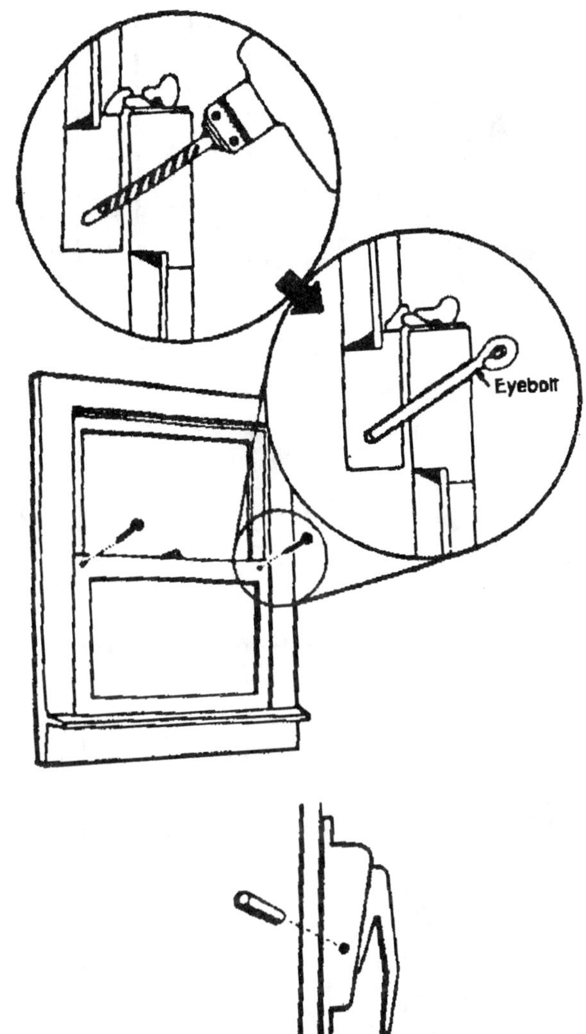

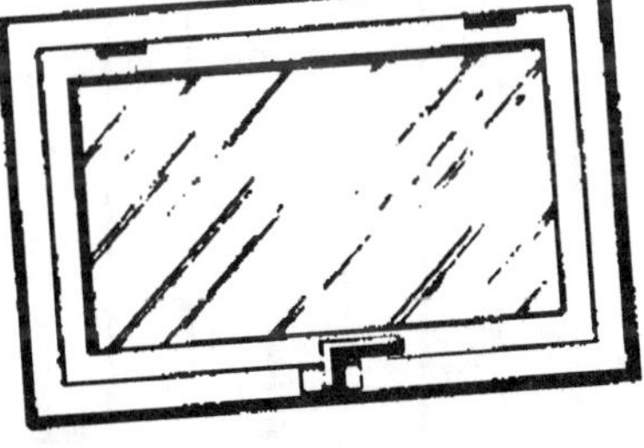

Casement windows are hinged at side, top, or bottom.

Some window latches can be secured with a drilled hole and inserted pin.

Keyed latch secures casement windows.

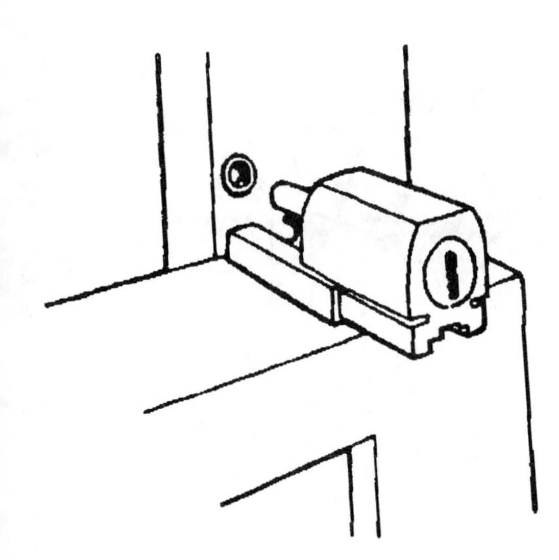

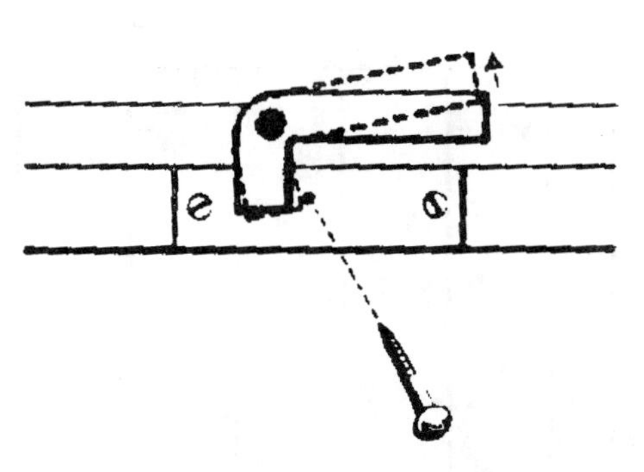

Screw prevents casement latch from opening.

Description of Suspect

Sex	Race	Age

Height	Weight	Complexion

Hair/Length Color
Glasses (type)
Tattoos/Scars/Mark
Facial Hair
Type Weapon

Hat (color, type)
Tie
Coat
Shirt
Pants/Shoes

Facial Appearance

Hair Style
Shape eyebrow
Sideburns
Shape and size of ears
Cheeks (full or sunken)
Beard/mustache
Chin clefts
Neck and adams apple
Hair texture
Wrinkles
Eye color
Nose shape
Mouth and lips

Write below specific facial details
—only what you definitely remember.

____

____

____

____

____

What did suspect say? ______________________________

____

____

Which way did he go? ______________________________

____

____

Auto Description Guide

Give this report to the first Police Officer on the scene.

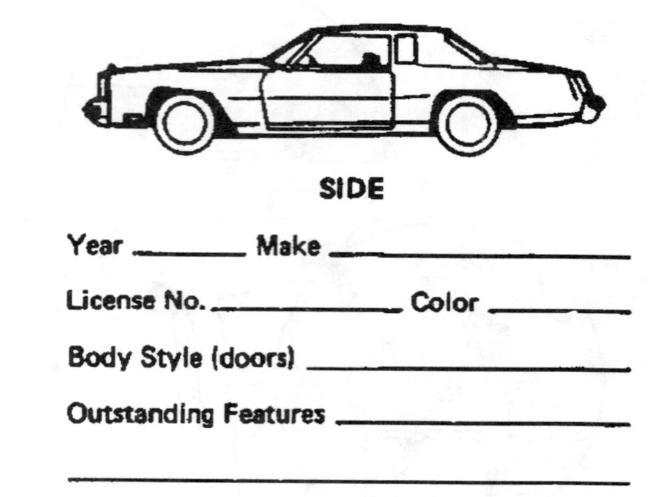

SIDE

Year ________ Make ______________

License No. ______________ Color ________

Body Style (doors) ______________

Outstanding Features ______________

____

____

FRONT REAR

No. Headlights ______ Shape Tailights ______

Which direction did car leave in? ______________

____

____

9-1-1 • Stay on the Telephone • Don't Hang Up

Chart 1.
A General View of The Criminal Justice System

This chart seeks to present a simple yet comprehensive view of the movement of cases through the criminal justice system. Procedures in individual jurisdictions may vary from the pattern shown here. The differing weights of line indicate the relative volumes of cases disposed of at various points in the system, but this is only suggestive since no nationwide data of this sort exists.

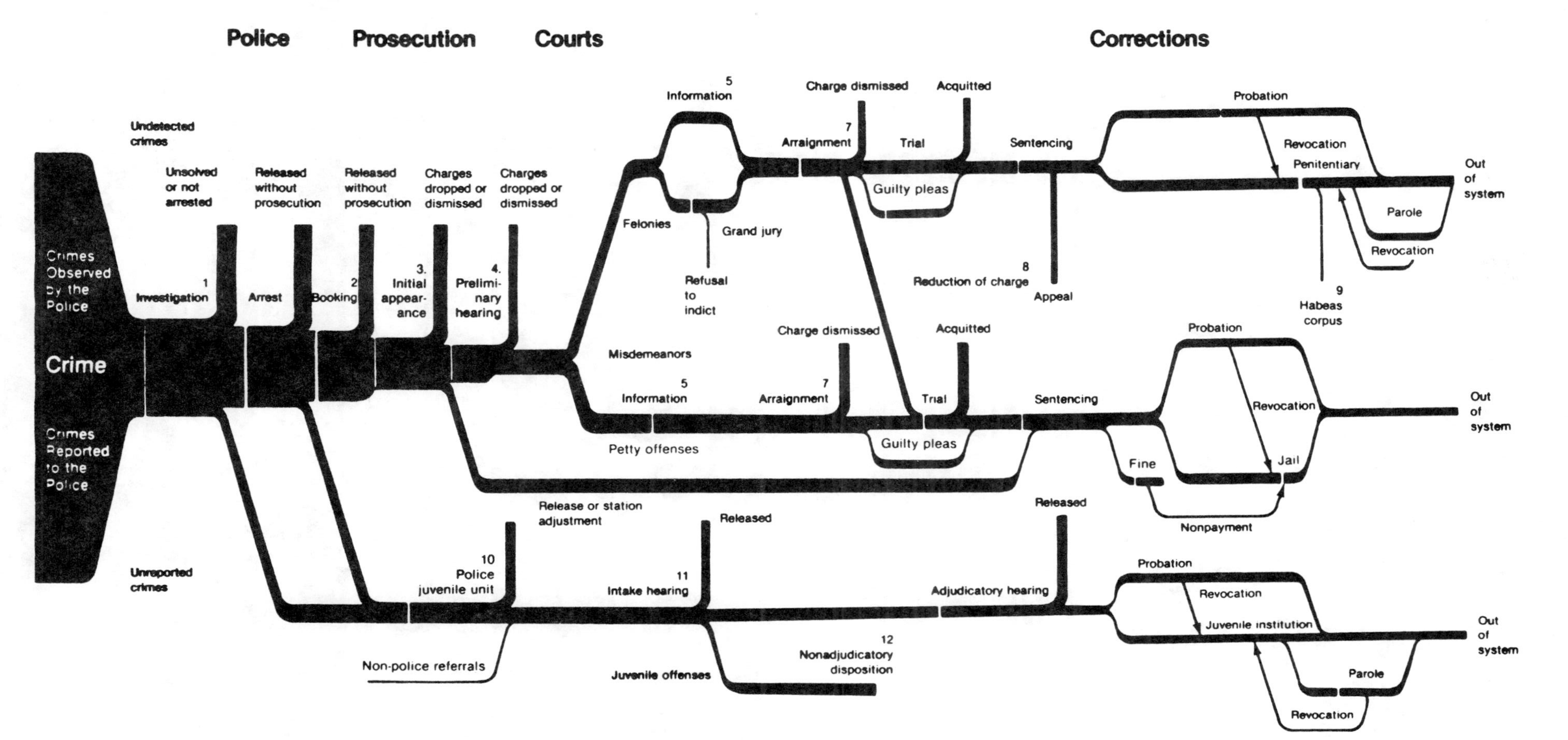

1. **May continue until trial.**
2. **Administrative record of arrest. First step at which temporary release on bail may be available.**
3. **Before magistrate**, commissioner, or justice of peace. Formal notice of charge, advise of rights. Bail set. Summary trials for petty offenses usually conducted here without further processing.
4. Preliminary testing of evidence against defendant. Charge may be reduced. No separate preliminary hearing for misdemeanors in same systems.
5. Charge filed by prosecutor on basis of information submitted by police or citizens. Alternative to grand jury indictment; often used in felonies, almost always in misdemeanors.
6. Reviews whether Government evidence sufficient to justify trial. Some States have no grand jury system, others seldom use it.
7. Appearance for plea; defendant elects trial by judge or jury (if available); counsel for indigent usually appointed here in felonies. Often not at all in other cases.
8. Charge may be reduced at any time prior to trial in return for plea of guilty or for other reasons.
9. Challenge on constitutional grounds to legality of detention. May be sought at any point in process.
10. Police often hold informal hearings, dismiss or adjust many cases without further processing.
11. Probation officer decides desirability of further court action.
12. Welfare agency, social services, counseling, medical care, etc., for cases where adjudicatory handling not needed.

In a major disaster, phone lines to emergency medical services may be overloaded or damaged. These pages will help you know what to do until medical help is available.

A sudden illness or physical injury can strike anyone at any time. More than 100,000 Americans die from accidents each year. 10,000,000 suffer disabling injuries. Medical authorities state that an alarming number of these people die or are disabled needlessly for lack of proper care immediately after the accident or at the start of the illness. They suggest that you carefully read the following pages and also take a first aid course from the American Red Cross.

When a person stops breathing death may occur in 4 to 6 minutes. When a person is bleeding badly, unless the bleeding is stopped within a few minutes the victim may die.
Remember: In an emergency, seconds and minutes can make the difference between life and death. Decisive, quick and proper action by you can save a life!

The Call for Help

1. If an injured person is in distress but is breathing...phone for help at once!

2. If the victim is not breathing...help first, and phone later...or get someone else to phone.

3. What to say:

A. Give the phone number from which you are calling.
B. Give the address and any special description of how to get to the victim.
C. Describe the victim's condition as best you can...burned, bleeding, broken bones...etc.
D. Give your name.
E. Do not hang up! Let emergency persons end the conversation. They may have questions to ask you or special information to give you about what you can do until help arrives.

Pacific Bell gratefully acknowledges the cooperation of the following agencies for their assistance in compiling and reviewing the procedures contained in Survival Guide.

American Red Cross, American Trauma Society, California Medical Association, Office of Emergency Services, Seismic Safety Commission

Emergency information regarding:

Breathing

Choking

Anything stuck in the throat blocking the air passage can stop breathing and cause unconsciousness and death within 4 to 6 minutes.

1. Do not interfere with a choking victim who can speak, cough or breathe. However, if the choking continues without lessening, call for emergency medical help.

2. If the victim cannot speak, cough or breathe, immediately have someone call for emergency medical help while you take the following action:

A. For a conscious victim:

1. Stand just behind and to the side of the victim who can be standing or sitting. Support the victim with one hand on the chest. The victim's head should be lowered. Give 4 sharp blows between the shoulder blades. If unsuccessful –

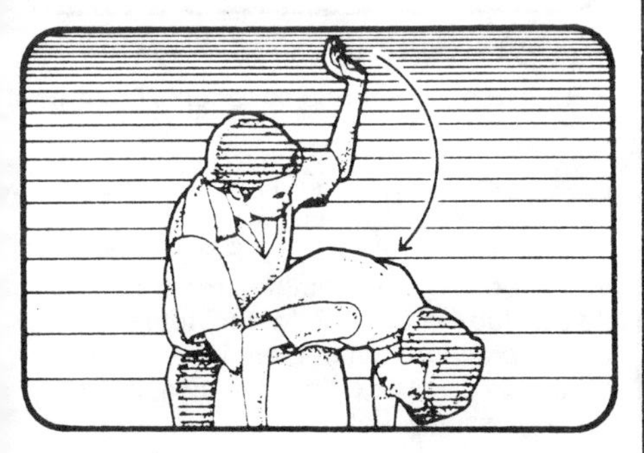

2. Stand behind the victim, who can be standing or sitting, wrap your arms around his or her middle just above the navel. Clasp your hands together in a doubled fist and press in and up in quick thrusts. Repeat several times.

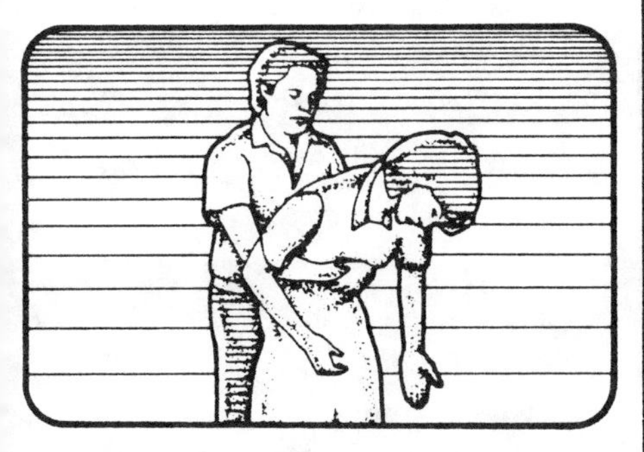

If still unsuccessful – Repeat 4 back blows and 4 quick thrusts until the victim is no longer choking or becomes unconscious.

B. For an unconscious victim:

1. Place the victim on the floor or ground and give rescue breathing. (See Rescue Breathing section.) If the victim does not start breathing and it appears that your air is not going into the victim's lungs –

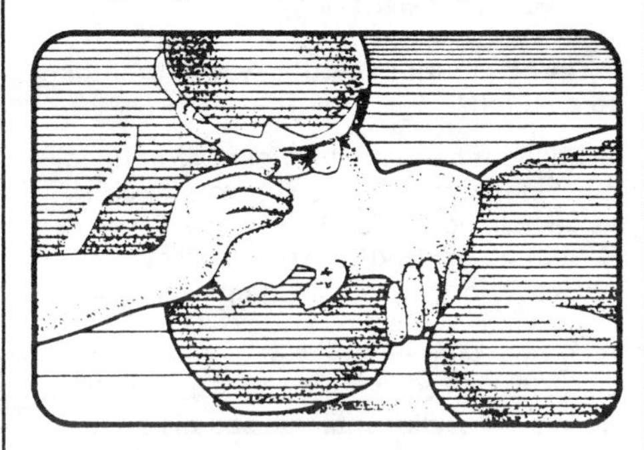

2. Roll the victim onto his/her side, facing you, with the victim's chest against your knee and give 4 sharp blows between the shoulder blades. If the victim still does not start breathing –

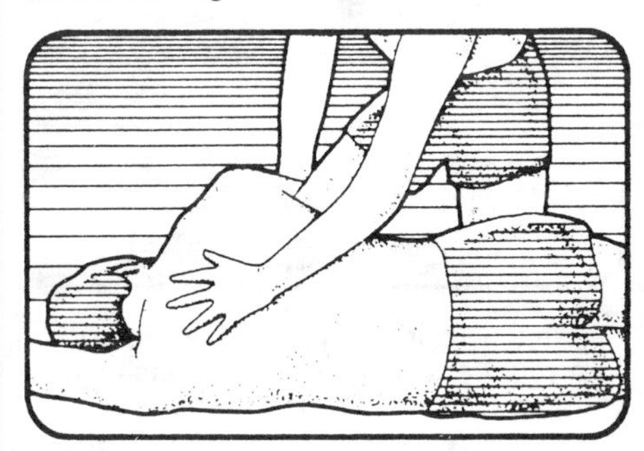

3. Roll the victim onto his or her back and give one or more manual thrusts. To give the thrusts, place one of your hands on top of the other with the heel of the bottom hand in the middle of the abdomen, slightly above the navel and below the rib cage. Press into the victim's abdomen with a quick upward thrust. Do not press to either side. Repeat 4 times if needed.

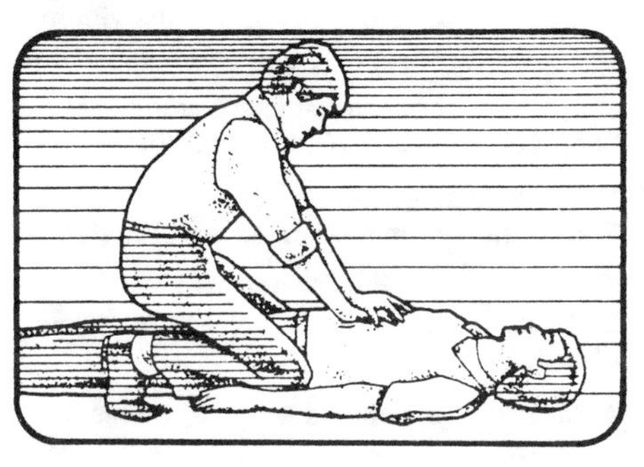

4. Clear the airway.
A. Hold the victim's mouth open with one hand using your thumb to depress the tongue.
B. Make a hook with the pointer finger of your other hand, and in a gentle sweeping motion reach into the victim's throat and feel for a swallowed foreign object which may be blocking the air passage. Repeat until successful:
(1) 4 back blows.
(2) 4 abdominal thrusts.
(3) Probe in mouth.
(4) Try to inflate lungs.
(5) Repeat

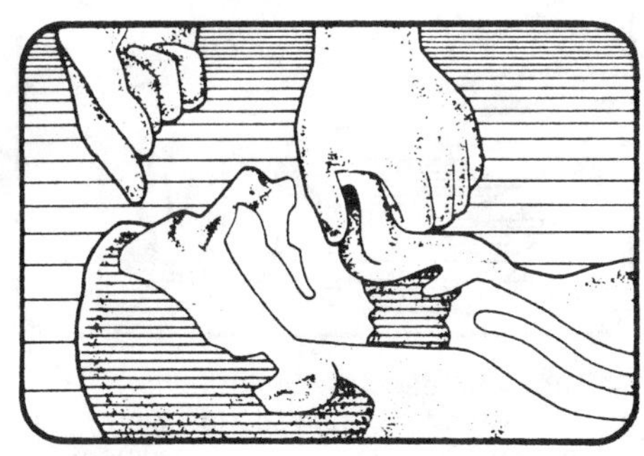

Note: If the object has not been retrieved, but the swallower suddenly seems all right, play it safe...take him or her directly to the hospital. This is especially critical if the swallowed object is a fish bone, chicken bone or other jagged object that could do internal damage as it passes through the victim's system.

continued ▶

Breathing

Unconscious Person

Breathing is the most critical thing we must do to stay alive. A primary cause of death is lack of air!

Be careful approaching an unconscious person. He or she may be in contact with electrical current. If that is the case, turn off the electricity before you touch the victim.

There are hundreds of other possible causes of unconsciousness, but the first thing you must check for is breathing.

1. Try to awaken the person: Shake the victim's shoulder vigorously. Shout: "Are you all right?"

2. If there is no response, check for signs of breathing.

A. Be sure the victim is lying flat on his or her back. If you have to, roll the victim over. Turn his or her head with remainder of body as a unit to avoid possible neck injury.
B. Loosen tight clothing around the neck and chest.

3. Open the airway:

A. If there are no signs of head or neck injuries, tilt the neck gently with one hand, causing the chin to protrude upward.
B. Push down and back on the forehead with the other hand as you tip the head back.

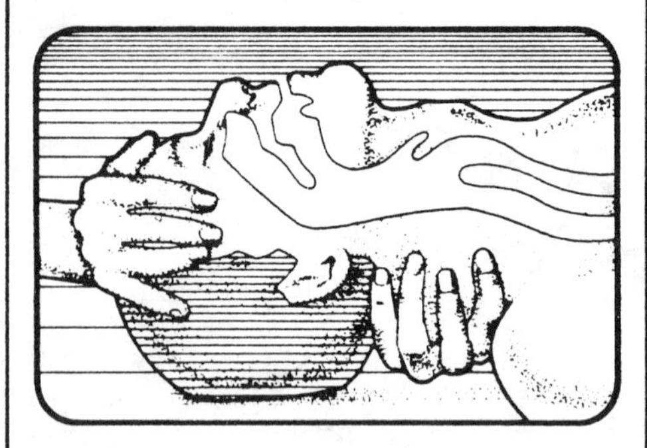

C. Place your ear close to the victim's mouth. Listen for breath sounds. Watch his or her chest and stomach for movement. Check for at least 5 seconds.

D. If there is any question in your mind, or if breathing is so faint that you are unsure... assume the worst!
E. Give rescue breathing immediately. Have someone else summon professional help.

Rescue Breathing

1. Giving mouth to mouth rescue breathing to an adult.

A. Put your hand on the victim's forehead, pinching the nose shut with your fingers, while holding the forehead back.
B. Your other hand is under the victim's neck supporting and lifting up slightly to maintain an open airway.
C. Take a deep breath. Open your mouth wide. Place it over the victim's mouth. Blow air into the victim until you see his or her chest rise.

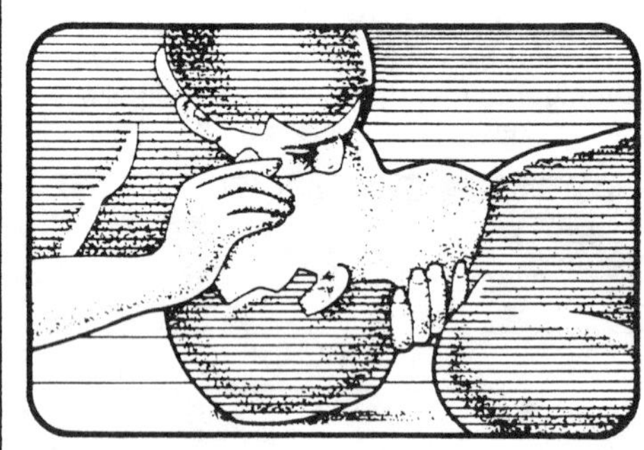

D. Remove your mouth from the victim's. Turn your head to the side and watch the chest for a falling movement while you listen for air escaping from the victim's mouth as he or she exhales.
E. If you hear air escaping and see the chest fall you know that rescue breathing is working. Continue until help arrives.
F. Initially give 4 breaths in rapid succession, then repeat single breath every 5 seconds. (12 breaths per minute.)

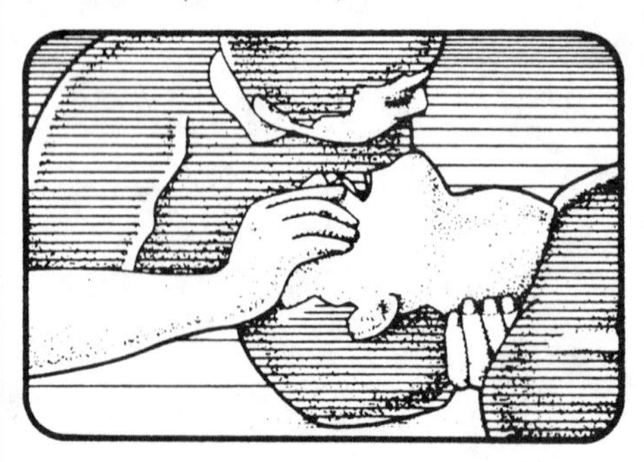

2. Giving mouth to mouth rescue breathing to infants and small children.

A. Be careful tilting a small child's head back to clear the airway. It should not be tilted as far back as an adult's. If tilted back too far, it will make the obstruction worse.
B. Cover the child's mouth and nose with your mouth.
C. Blow air in with less pressure than for an adult. Give small puffs. A child needs less.
D. Feel the chest inflate as you blow.

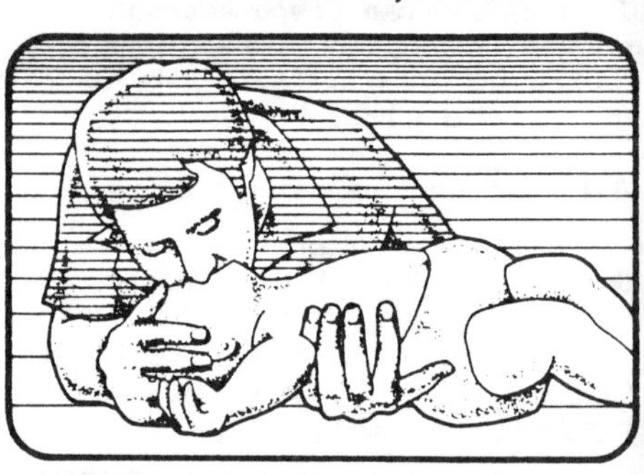

E. Listen for exhales.
F. Repeat once every 3 seconds. 20 breaths per minute.

Note: It may take several hours to revive someone. Keep up rescue breathing until help arrives to relieve you. Remember, you are doing the breathing for the victim. If you stop – in about 4 to 6 minutes – he or she could be dead! Even if the victim begins to breathe on his/her own, call for professional help.

Breathing and Electric Shock

Drowning

Drowning is a major cause of accidental death in the United States. Victims who die of drowning can die within about 4 to 6 minutes of the accident because they have stopped breathing.

1. Get the victim out of the water at once.

Use extreme caution to avoid direct contact with the victim since a panicked victim may drown the rescuer as well.

If the victim is conscious, push a floating object to him/her or let the victim grasp a long branch, pole or object.

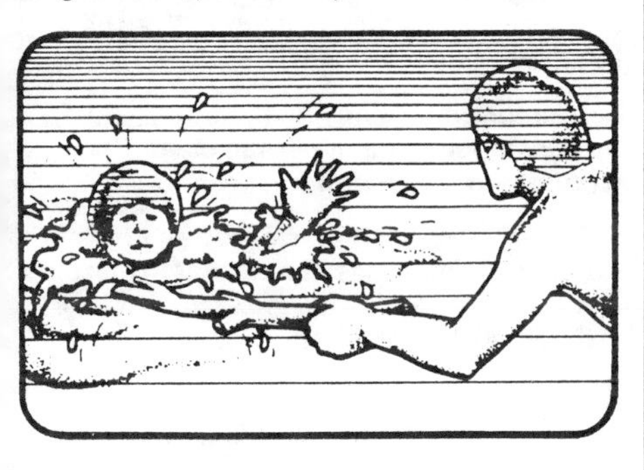

If the victim is unconscious, take a floatation device with you if possible and approach the victim with caution. Once ashore or on the deck of a pool, the victim should be placed on his/her back.

2. If the victim is not breathing, start mouth to mouth rescue breathing immediately. (See Rescue Breathing section.)

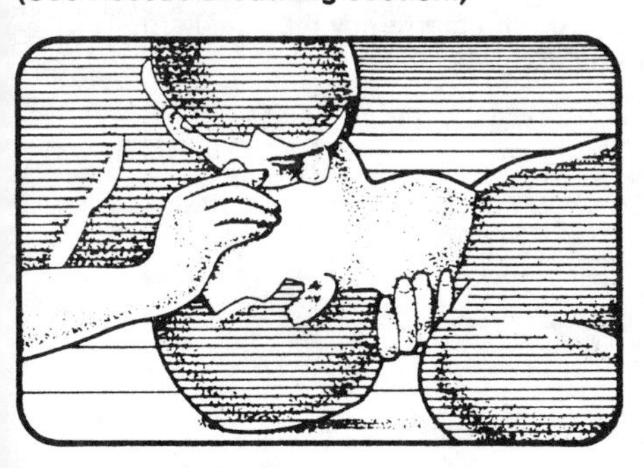

Keep giving rescue breathing until the victim can breathe unassisted. That can take an hour or two. Pace yourself. Keep calm. Remember: Even when the victim is breathing unassisted, he or she may be in need of medical attention. Have someone else go for help. Do not leave the victim alone under any circumstances...not even to call for help!

3. If the victim is breathing without assistance, even though coughing and sputtering, he or she will get rid of the remaining water. You need only stand by to see that recovery continues, but have someone else send for professional help.

Electric Shock

(Electrocution) Normal electrical current can be deadly, and it is all around us.

1. Do not touch a person who has been in contact with electrical current until you are certain that the electricity has been turned off. Shut off the power at the plug, circuit breaker or fuse box.

2. If the victim is in contact with a wire or a downed power line, use a dry stick to move it away.

3. Check for breathing – If the victim's breathing is weak or has stopped:

A. Give Rescue Breathing immediately. (See Rescue Breathing section.)

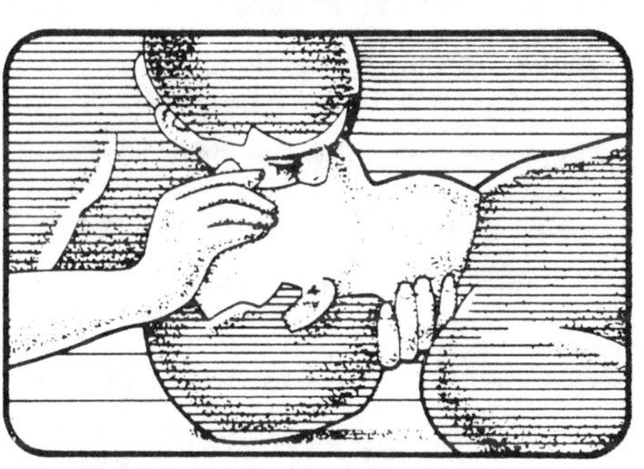

4. Call for emergency help. While you wait for help to arrive:

A. Keep the victim warm (covered with a blanket, coat, jacket, etc.).
B. Give the victim nothing to drink or eat, until he or she is seen by a doctor.

continued ▶

Heart Attack and Bleeding

Heart Attack

Heart attack is the number one killer of adults over the age of 38. Many heart attack victims die needlessly because they do not get help in time.

1. Warning signs include:

A. Severe squeezing pains in the chest.
B. Pain that radiates from the chest into either the arm, the neck or jaw.
C. Sweating and weakness, nausea or vomiting.
D. Pain that extends across the shoulders to the back.

2. If the victim is experiencing any of these sensations take no chances. Call for emergency help at once.

3. Two critical life threatening things happen to the victim of a heart attack:

A. Breathing slows down or stops.
B. The heart may slow down or stop pumping blood.

4. If the victim is not breathing:

A. Give rescue breathing immediately and have someone else call for emergency help.

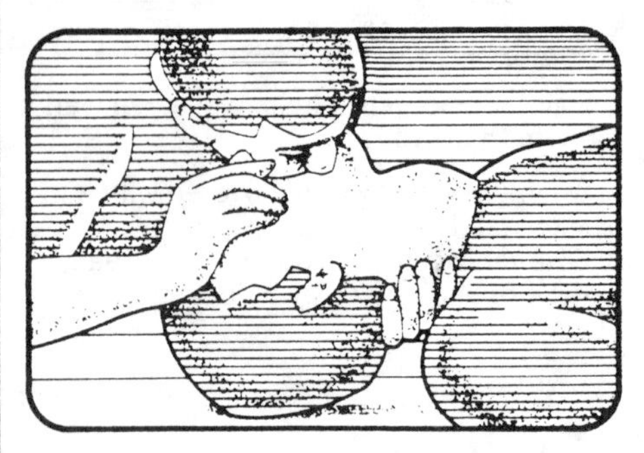

5. If you cannot detect a heart beat by taking a pulse at the Carotid Artery:

(The Carotid Artery can be felt on either side of the neck slightly below and forward of the base of the jaw).

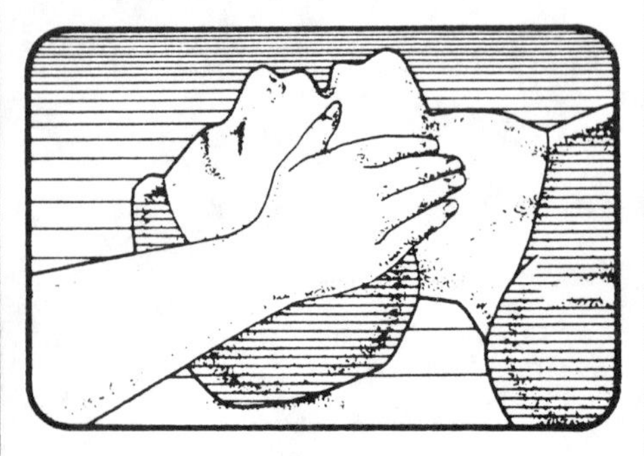

A. "CPR", Cardiopulmonary Resuscitation should be given to the victim along with Rescue Breathing only by a person properly trained and certified.

Learn CPR. CPR is a way of forcing the heart to continue pumping blood (carrying oxygen) through the lungs and out to the rest of the body where it is needed if life is to continue. CPR is too complicated to be taught from the printed pages alone. Courses are offered by The American Heart Association and The American Red Cross. Many medical authorities agree that every-one thirteen years of age and older should learn both CPR and Rescue Breathing.

Bleeding

Check to see if the victim is wearing a Medic Alert or similar bracelet, necklace, etc. It describes emergency medical requirements.

The best way to control bleeding is with direct pressure over the site of the wound.

A. Use a pad of sterile gauze, if one is available.
B. A sanitary napkin, a clean handkerchief, or even your bare hand, if necessary, will do.
C. Apply firm, steady direct pressure for 5 to 15 minutes. Most bleeding will stop within a few minutes.

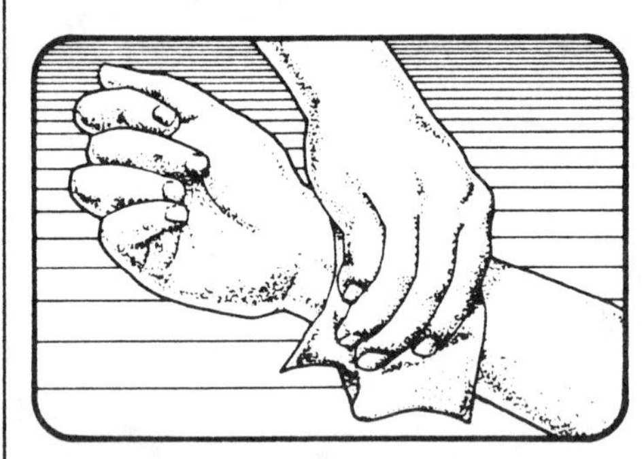

D. If bleeding is from a foot, hand, leg or arm use gravity to help slow the flow of blood. Elevate the limb so that it is higher off the ground than the victim's heart.
E. Severe nose bleeding can often be controlled by direct pressure such as by pinching the nose with the fingers. Apply pressure 10 minutes without interruption.

Head Injuries

If there is bleeding from an ear, it can mean that there is a skull fracture.

1. Special care must be taken when trying to stop any scalp bleeding when there is a suspected skull fracture. Bleeding from the scalp can be very heavy even when the injury is not too serious.

2. Don't press too hard. Be extremely careful when applying pressure over the wound so that bone chips from a possible fracture will not be pressed into the brain.

3. Always suspect a neck injury when there is a serious head injury. Immobilize the head and neck.

4. Call for emergency help. Let a professional medical person clean the wound and stitch it, if necessary.

5. Do not give alcohol, cigarettes or other drugs. They may mask important symptoms.

Internal Bleeding

Warning Signs: coughing or vomiting up blood or "coffee ground" material. Passing blood in urine or stool. Passing black tar-like bowel movements. All require immediate medical attention!

1. Have the victim lie flat and relax.

2. Do not let the victim take any medication or fluids by mouth until seen by a doctor who permits it.

3. Obtain emergency medical help immediately.

Broken Bones and Seizure

Broken Bones

Broken bones usually do not kill. Do not move the victim, unless the victim is in immediate danger of further injury.

1. Check for:

A. Breathing Give Rescue Breathing if needed.

B. Bleeding Apply direct pressure over the site.

C. Shock Keep the victim calm and warm.

2. Call for emergency help.

3. Do not try to push the broken bone back into place if it is sticking out of the skin. Do apply a moist dressing to prevent drying out.

4. Do not try to straighten out a fracture. Let a doctor or trained person do that.

5. Do not permit the victim to walk about.

6. Splint unstable fractures to prevent painful motion.

Seizure

It is an alarming sight; a person whose limbs jerk violently, whose eyes may roll upward, whose breath may become heavy with dribbling or even frothing at the mouth. Breathing may stop in some seizures, or the victim may bite his or her tongue so severely that it may bleed and cause an airway obstruction. Do not attempt to force anything into the victim's mouth. You may injure the victim and yourself.

1. During the seizure:

A. There is little you can do to stop the seizure.
B. Call for help.
C. Let the seizure run its course.
D. Help the victim to lie down and keep from falling, to avoid injury.
E. Loosen restrictive clothing.
F. Use no force.
G. Do not try to restrain a seizure victim.
H. Move objects out of the way which may injure the victim (i.e. desk, table, chair, etc.).
I. If an object endangers the victim and cannot be moved, put clothing or soft material between the seizure victim and the object.

2. After the seizure:

A. Check to see if the victim is breathing... If he or she is not...give Rescue Breathing at once. (See Rescue Breathing section.)
B. Check to see if the victim is wearing a Medic Alert, or similar, bracelet, necklace, etc. It describes emergency medical requirements.

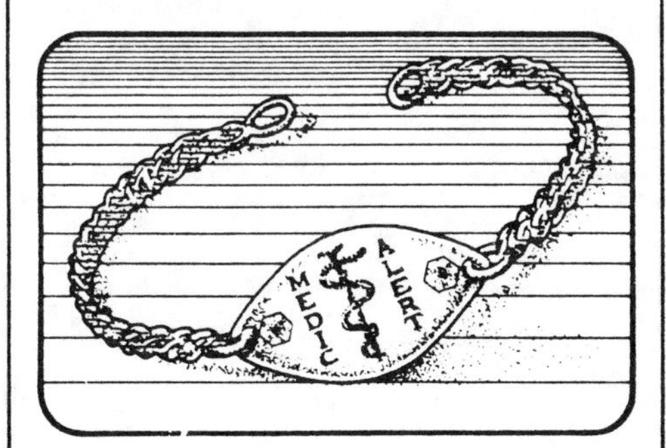

C. Check to see if the victim has any burns around the mouth. This would indicate poison.

3. The victim of a seizure or convulsion may be conscious, but confused and not talkative when the intense movement stops. Stay with the victim. Be certain that breathing continues. Then, when the victim seems able to move, get medical attention.

continued ▶

Poisoning and Burns

Poisoning

The home is loaded with poisons: Cosmetics, Detergents, Bleaches, Cleaning Solutions, Glue, Lye, Paint, Turpentine, Kerosene, Gasoline and other petroleum products, Alcohol, Aspirin and other medications, and on and on.

1. Small children are most often the victims of accidental poisoning. If a child has swallowed or is suspected to have swallowed any substance that might be poisonous, assume the worst - Take Action.

2. Call your Poison Control Center. If none is in your area, call your emergency medical rescue squad. Bring suspected item and container with you.

A. Do **not** give counteragents unless directed to by Poison Control Center or physician.
B. Do **not** follow directions for neutralizing poisons found on poison container.
C. If victim is conscious, dilute poison by giving moderate amounts of water.

3. What you can do if the victim is unconscious:

A. Make sure patient is breathing. If not, tilt head back and perform mouth to nose breathing. Do not give anything by mouth. Do not attempt to stimulate person. Call emergency rescue squad immediately.

4. If the victim is vomiting:

A. Roll him or her over onto the left side so that the person will not choke on what is brought up.

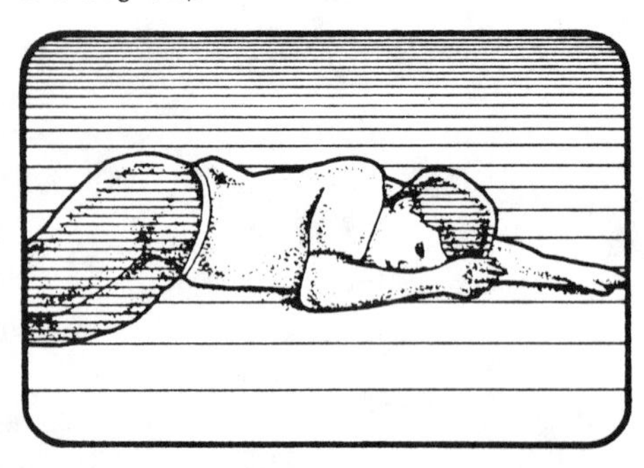

5. Be prepared. Determine and verify your Poison Control Center and Fire Department Rescue Squad numbers and keep them near your telephone.

Drug Overdose

A drug overdose is a poisoning. Alcohol is as much a poison as stimulants, tranquilizers, narcotics, hallucinogens or inhalants. Don't take drunkenness lightly. Too much alcohol can kill.

1. Call for emergency help at once.

2. Check the victim's breathing and pulse. If breathing has stopped or is very weak give Rescue Breathing. Caution: Victims being revived of alcohol poisoning can be violent. Be careful! They can harm themselves and others.

3. While waiting for help:

A. Watch breathing.
B. Cover the person with a blanket for warmth.
C. Do not throw water on the victim's face.
D. Do not give liquor or a stimulant.

Remember: alcohol in combination with certain other drugs can be deadly!

Burns

Flame Burns
Cool with water to stop the burning process.

Remove garments and jewelry and cover burn victim with clean sheets or towels.

Call for help immediately.

Chemical Burns
Remove clothing.

Wash with **cool water** for at least **20 minutes.**

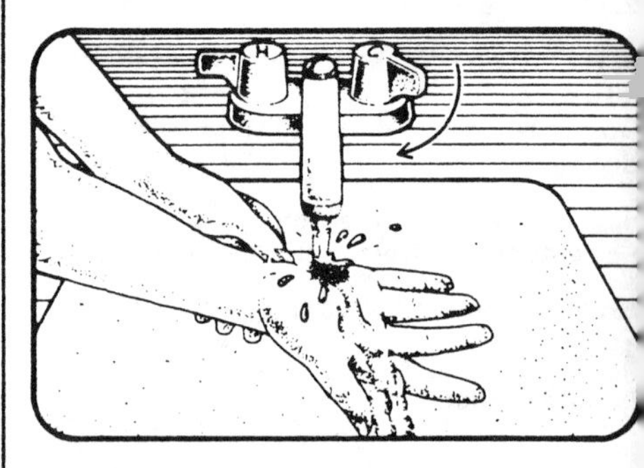

Call for help immediately.

Chemical burns of the **eye** require immediate medical attention after flushing with water for 20 minutes.

Earthquake – Before and During

'here will always be earthquakes in California. Scientists are working on a ong range goal of earthquake prediction, but even if they succeed there will always be a need for citizen preparedness...and he ability of individuals to take care of hemselves and their loved ones in time of emergency. Know the location of your nearest Fire and Police Station.

. Before an earthquake happens, be prepared. Have basic emergency supplies on hand:

A. A portable radio (with extra batteries).
B. A few flashlights (with extra batteries).
Note: Batteries last longer if stored in a refrigerator.
C. A first aid kit and handbook.
D. Water (a few gallons for each family member).
E. Food (canned foods, mechanical opener, equired medications and powdered milk or at least one week's meals).
F. Pipe wrenches and crescent wrenches to be able to turn off gas and water).
G. Know where your gas, electric and water main shutoffs are. If in doubt, ask your water, power, and gas companies.
H. Have some alternate source for cooking which can be used outdoors (i.e., barbecue, charcoal, starter fluid, matches – the latter wo items stored separately and out of the each of children).
. A small bottle of chlorine bleach for use n disinfecting water.
. Have a plan to reunite your family since ravel may be difficult or even restricted after a major earthquake.

2. How to shut off gas:
(Do so only if you suspect a gas leak or can smell escaping gas.)

A. The main shut-off valve is located next to your meter on the inlet pipe.
B. Use a wrench and give it a quarter turn in either direction so that it runs cross wise on the pipe. The line is now closed.

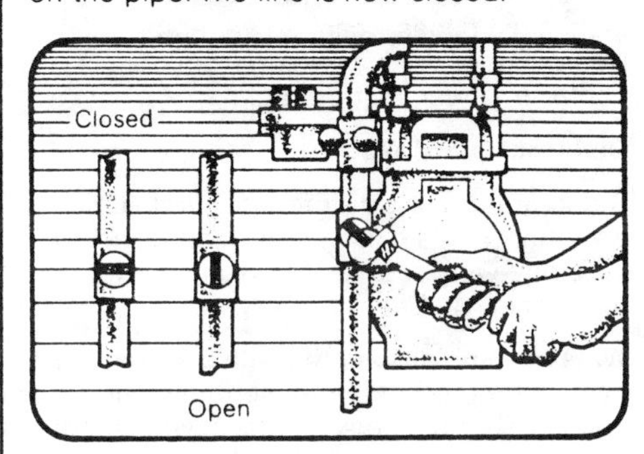

3. How to shut off electricity:

A. Look closely at your circuit breaker box or fuse-type box.
B. Be certain that you can turn off the electricity in an emergency.

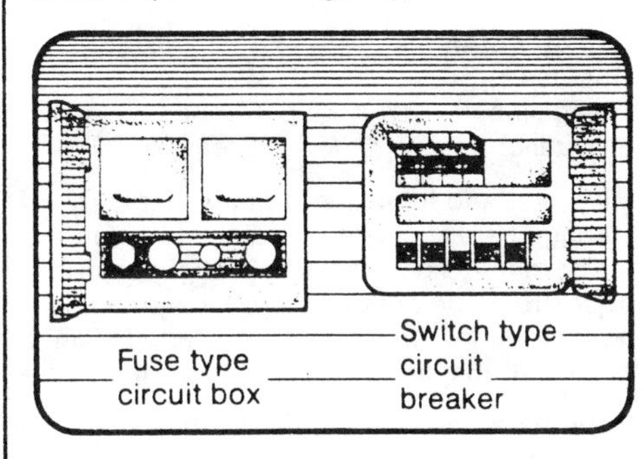

4. During an earthquake, keep calm. Panic kills.

A. If you are indoors, stay there. Get under a desk, table or in a doorway. Stay clear of windows. Greatest hazards from falling objects may be outdoors – Stay inside!
B. If you are outside – Get into the open, away from buildings and power lines.
C. If you are driving a car – Stop the car, but stay inside. Do not stop on or under a bridge. Try to get your car as far out of the normal traffic pattern as possible. Do not stop under trees, lightposts or signs.
D. If you are in a mountainous area, be alert for falling rock or other debris which could be loosened as a result of the quake.

continued ▶

Earthquake – After

5. After a major earthquake:

A. Check For Injuries – If anyone has stopped breathing, give mouth to mouth Rescue Breathing. · Stop any bleeding injury by applying direct pressure over the site of the wound. · Do not attempt to move seriously injured persons unless they are in immediate danger of further injury. · Cover injured persons with blankets to keep them warm. Be reassuring and calm. · Wear shoes in all areas near debris and broken glass. · Immediately clean up any spilled medicines, drugs or other potentially harmful materials (e.g. bleaches, lye, gasoline, or other petroleum products).
B. Check For Safety – Check your home for fire or fire hazards. · Check utility lines and appliances for damage. · Shut off main gas valve. Do not search for a leak with a match. Do not turn on gas again. You will be safer if you let the gas company restore service. · Shut off electrical power at the control box if there is any damage to your house wiring. · Do not use lighters or open flame appliances until you are certain that no gas leak exists. · Do not operate electrical switches or appliances if gas leaks are suspected. Sparks can ignite gas from broken lines. · Do not touch downed lines, or objects touched by downed power lines, or electrical wiring of any kind. · Check your chimney for cracks and damage. Approach chimneys with caution. They may topple. Caution: Use of a damaged chimney invites fire. When in doubt, don't use it. · Check closets and cupboards Open doors cautiously. Beware of falling objects tumbling off shelves.
C. Check Your Food Supply – Do not eat or drink anything from open containers near shattered glass. If power is off, check your freezer and plan meals to use up foods that will spoil quickly. Use outdoor charcoal broilers for emergency cooking.
D. Check your water supply – If water is off, emergency water supplies may be all around you in...water heaters...toilet tanks... melted ice cubes...canned vegetables. Do not eat or drink anything from open containers near shattered glass.

Disinfection of Water

Before attempting disinfection, first strain water through a clean cloth or handkerchief to remove any sediment, floating matter or glass.

Water may be disinfected with 5.25% sodium hypochlorite solution (household chlorine bleach). Do not use solutions in which there are active ingredients other than hypochlorite. Use the following proportions:

Clear water	
One quart	2 drops
One gallon	8 drops
Five gallons	½ teaspoon

Cloudy water	
One quart	4 drops
One gallon	16 drops
Five gallons	1 teaspoon

Mix water and hypochlorite thoroughly by stirring or shaking in a container. Let stand for 30 minutes before using. A slight chlorine odor should be detectable in the water; if not, repeat the dosage and let stand for an additional 15 minutes before using.

Note: Water may be purified by bringing it to a rapid boil.

6. Cooperate with Public Safety efforts.

A. Do not use your telephone except to report medical, fire or violent crime emergencies.
B. Turn on your portable radio for information and damage reports.
C. Do not go sightseeing afterwards, especially in beach and waterfront areas where seismic waves could strike.
D. Keep streets clear for emergency vehicles.
E. Be prepared for aftershocks. Most of these are smaller than the main quake, but some may be large enough to do additional damage.
F. Cooperate with Public Safety Officials. Don't go into damaged areas unless your help is requested.
G. Informed and cooperative citizens can help minimize damage and injury.

INDEX